SpringerBriefs in Applied Sciences and Technology

SpringerBriefs present concise summaries of cutting-edge research and practical applications across a wide spectrum of fields. Featuring compact volumes of 50 to 125 pages, the series covers a range of content from professional to academic.

Typical publications can be:

- A timely report of state-of-the art methods
- An introduction to or a manual for the application of mathematical or computer techniques
- A bridge between new research results, as published in journal articles
- A snapshot of a hot or emerging topic
- An in-depth case study
- A presentation of core concepts that students must understand in order to make independent contributions

SpringerBriefs are characterized by fast, global electronic dissemination, standard publishing contracts, standardized manuscript preparation and formatting guidelines, and expedited production schedules.

On the one hand, **SpringerBriefs in Applied Sciences and Technology** are devoted to the publication of fundamentals and applications within the different classical engineering disciplines as well as in interdisciplinary fields that recently emerged between these areas. On the other hand, as the boundary separating fundamental research and applied technology is more and more dissolving, this series is particularly open to trans-disciplinary topics between fundamental science and engineering.

Indexed by EI-Compendex, SCOPUS and Springerlink.

More information about this series at https://link.springer.com/bookseries/8884

Pratima Bajpai

Fourth Generation Biofuels

 Springer

Pratima Bajpai
Pulp and Paper Consultant
Kanpur, India

ISSN 2191-530X   ISSN 2191-5318   (electronic)
SpringerBriefs in Applied Sciences and Technology
ISBN 978-981-19-2000-4   ISBN 978-981-19-2001-1   (eBook)
https://doi.org/10.1007/978-981-19-2001-1

This Springer imprint is published by the registered company Springer Nature Singapore Pte Ltd.
The registered company address is: 152 Beach Road, #21-01/04 Gateway East, Singapore 189721, Singapore

Preface

Growing concerns about the rapid depletion of fossil fuel reserves, rising crude oil prices, energy security, and global climate change have led to increased worldwide interest in renewable energy sources such as biofuels. In this context, biofuel production from renewable sources is considered to be one of the most sustainable alternatives to fossil fuels and a viable means of achieving environmental and economic sustainability.

Although biofuel processes hold great potential to provide a carbon-neutral route to fuel production, first-generation production systems are characterized by considerable economic and environmental limitations. The advent of second-generation biofuels is intended to produce fuels from lignocellulosic biomass, the woody part of plants that do not compete with food production. However, converting woody biomass into fermentable sugars requires costly technologies. Third-generation biofuels, which are derived from algae, have carbon dioxide (CO_2) and relatively simple processing. Algae can be cultivated in wastewater and seawater as well as in unproductive drylands and marginal farmlands. Thus, they do not compete with food crops on arable land or in freshwater environments. Fourth-generation biofuels aim at providing more sustainable production options by combining biofuels production with capturing and storing CO_2 with the process of oxy-fuel combustion or by application of genetic engineering or nanotechnology. The main feedstock for the fourth-generation biofuels production is genetically engineered, highly yielding biomass with low lignin and cellulose contents (thus eliminating the issues present in the second-generation biofuels production line) or metabolically engineered algae (with high oil contents, increased carbon entrapment ability, and improved cultivation, harvesting, and fermentation processes) (thus improving the third-generation production). Fourth-generation biofuels are thought to contribute better to reducing GHG (greenhouse gas) emissions, by being more carbon-neutral or even carbon-negative compared to the other generation biofuels. This book examines the background of fourth-generation biofuel production, use of genetically modified microalgae for the production of fourth-generation biofuels; cultivation and harvesting of genetically modified microalgae; residue from biofuel extraction; health and environmental concerns of fourth-generation biofuels; regulations on cultivation and processing of the genetically modified algae; carbon dioxide sequestration; water footprint; and current status and key challenges.

Kanpur, India   Pratima Bajpai

Acknowledgments

I am grateful for the help of many people, companies, and publishers for providing information and granting permission to use their material. Deepest appreciation is extended to Elsevier, Springer, Taylor and Francis Sons, Hindawi, MDPI, IntechOpen, and other open-access journals and publications. My special thanks to Dr. Shashi Kumar, International Centre for Genetic Engineering and Biotechnology, New Delhi, India for providing information on GM Algae for Biofuel Production.

Contents

List of Figures

List of Tables

Chapter 1
Introduction and General Background

Since the early 2000s, a combination of surging energy demand and over-utilization of natural resources has resulted in a sharp, alarming exhaustion of fossil fuels. Fossil fuels have taken millions of years to get formed and are non-renewable resources. Fossil fuel sources are getting exhausted and contribute to emissions of greenhouse gases leading to several unfavorable effects which are change in climate, receding of glaciers, increase in sea level, and loss of biodiversity. Moreover, experts have calculated that the earth's fossil fuel reserves will be totally diminished in forty years' time if we continue to be over-dependent on them (Shafiee and Topal 2009).

"Biofuels are widely lauded as a key solution to the world's energy crisis. Ironically, biofuels had long existed before the discovery of fossil fuels; however, they were essentially disregarded due to the large reserves of gas, coal, and crude oil available at the time; this in conjuncture with the relatively low costs of fossil fuels made them highly desirable in the past, especially in developed countries. Today, the thinning supplies of these conventional fuels have sparked the search for an alternative, renewable answer to meeting our energy needs. Therefore, interest in the development of biofuels, and their uses have been reignited. Biofuels are viewed on a global scale to have an immediate impact on plaguing concerns such as escalating oil prices and emission of greenhouse gases. While biofuels are commonly referred to as the fuel of the future, the idea was first conceptualized and developed by Rudolph Diesel in the late nineteenth century. Since its inception, biofuels have been used primarily in the automotive industry due to their potential to replace gasoline and diesel. However, advancements in the field have shown that aside from its utility as a sustainable transport fuel, biofuel can also be used for manufacturing, cosmetic, pharmaceutical, heating, and agricultural processes. Nevertheless, there have been ongoing debates concerning the use of agrofuels over fossil fuels despite the former's obvious market potential and environmental benefits. The skeptics cited issues regarding the economic and environmental impacts of producing biofuel, raising questions over its feasibility and sustainability. In spite of this, continued research and development

© The Author(s), under exclusive license to Springer Nature Singapore Pte Ltd. 2022   1
P. Bajpai, *Fourth Generation Biofuels*,
SpringerBriefs in Applied Sciences and Technology,
https://doi.org/10.1007/978-981-19-2001-1_1

Table 1.1 Types of biofuels substitute for fossil fuel

Types of biofuel		
Biodiesel	Bio-oil	Bioethanol
Diesel substituent	Kerosene substituent and fuel oil	Gasoline substituent

have been invested in the enhancement of biofuels, negating most of the abovementioned concerns. Additionally, governments, industry players, and civil society have started several initiatives to develop criteria for sustainable production of biofuels (United Nations Environmental Programme 2009). Through the combined efforts of meticulous administration, as well as the constant endeavor of improving biofuels, the merits of pursuing a future powered by biofuel will far outweigh any consequences eventually" (Olaganathan et al. 2014).

There are different types of biofuels available for commercial use. They can be classified as biodiesel, bioethanol, and bio-oils as depicted in Table 1.1.

Biofuels may be derived from agricultural crops, including conventional food plants, or from special energy crops. Biofuels may also be derived from forestry, agricultural, or fishery products or municipal wastes, as well as from agro-industry, food industry, and food service by-products and wastes. A distinction is made between primary and secondary biofuels (Table 1.2). In the case of primary biofuels, such as fuelwood, wood chips, and pellets, organic materials are used in an unprocessed form, primarily for heating, cooking, or electricity production. Secondary biofuels result from the processing of biomass and include liquid biofuels such as ethanol and biodiesel that can be used in vehicles and industrial processes.

Biofuels are made from biomass through processes such as chemical, biochemical, or hybrid conversions (Amin 2009). Biofuels do not create additional emissions apart from those produced during the production and transportation stages (Lee and Lavoie 2013). Hence, biofuels avoid the environmental drawbacks associated with the consumption of fossil fuels. Biofuels are usually classified into four classes as first-, second-, third-, and fourth-generation biofuel (Table 1.2).

First-generation biofuel includes bioethanol and biodiesel produced from sugar and starch crops and oil, respectively. Bioethanol is the most widely used biofuel for transportation worldwide (Balat 2011; Bajpai 2013; Cesaro and Belgiorno 2015; Clausen and Gaddy 2009; Huhnke 2009; Walker 2010; Farell et al. 2006; Sivakumar et al. 2010; Gnansounou 2010; Gomez et al. 2008; Lynd 1996; Wooley et al. 1999; Graf and Koehler 2000; Biofuels Digest 2008; IEA 2008; Mabee and Saddler 2007; Naik et al. 2010; Huber et al. 2006; Carroll and Somerville 2009; Pu et al. 2008; Ragauskas et al. 2006a, b).

First-generation bioethanol is mainly produced from food crop feedstock; thus, it is competing for agricultural areas used for food production. Hence, it is believed that the efficiency of the ethanol produced from the first-generation feedstock to achieve targets for the substitution of fossil fuels in reducing global warming and to reach economic growth is limited (Borse and Sheth 2017). The cumulative impact of these concerns has increased pressure on the energy sector in producing ethanol from the

Table 1.2 Biofuels

Primary biofuel
Firewood, wood chips, pellets, animal waste, forest and crop residues, landfill gas

Secondary fuel
First-generation biofuels
Produced from food crops (by fermentation of starch from wheat, barley, corn, potato or sugars from sugarcane, sugar beet, etc.)
(a) biodiesel extracted from oil plants/plant materials (in the chemical process of transesterification and esterification)
(b) ethanol extracted from sugar-containing plants/plant materials and converted to fuel in the process o fermentation
Second-generation biofuels
Produced from nonfood crops, e.g., crop waste, green waste, wood, and energy crops planted specifically for biofuels production
Third-generation biofuels
Biodiesel from microalgae
Bioethanol from microalgae and seaweeds
Hydrogen from green microalgae and microbes
Fourth-generation biofuels
Aims at providing more sustainable production options by combining biofuels production with capturing and storing carbon dioxide with the process of oxy-fuel combustion or by application
This carbon capture makes fourth-generation biofuel production carbon-negative rather than simply carbon–neutral, as it locks away more carbon than it produces
This system not only captures and stores carbon dioxide from the atmosphere but also reduces CO_2 emissions by replacing fossil fuels

Based on Ziolkowska (2018, 2020), Alam et al. (2012), Yanqun et al. (2008) Behera et al. (2014, 2015), Boyce et al. (2008), Shuping et al. (2010), Hughes et al. (2012), Singh et al. (2011, 2014); Ullah et al. (2018)

non-edible feedstock. In this regard, the second-generation biofuel was developed to overcome the deficiencies of the first-generation biofuel in terms of food security, biodiversity, and sustainability.

Second-generation biofuels differ in feedstock which, in this case, comes from nonfood plants such as agricultural crops, residues, and wood (so-called lignocellulosic biomass). Biorefinery is gaining wide attention for converting lignocellulosic biomass to biofuels. It is a concept analogous to petroleum refineries, which produce various output products from crude oil. The idea behind designing a biorefinery is to integrate different conversion techniques to produce a wide variety of products by taking advantage of all the fractions in the biomass (Aguilar et al. 2018). Biorefinery integrates producing bioethanol or other biochemicals sustainably from lignocellulosic biomass. Palm oil is one of the energy crops widely used for second-generation biofuel due to its high rate of productivity—per unit of planted area—compared to other oilseeds such as soybean and rapeseed. Deforestation and the associated health risks are the key concerns faced by governments as a result of the over-reliance on palm oil.

Third-generation biofuel refers to biofuel obtained from algae. The third-generation biofuels are more energy-dense in comparison to earlier generations (yield

Table 1.3 Main advantages of microalgae in comparison to superior plants

Faster growth, large carbon dioxide absorption capacity
Ability to tolerate large temperature variations resulting from daily and seasonal cycles
Low nutrient requirement
Potential production of high-value co-products in addition to the desired product
Short productive cycle
High photosynthetic efficiency
Ability to tolerate and adapt to a variety of environmental conditions
Ability for some species to grow heterotrophically, consuming (residual) organic carbon in industrial effluents

per area of harvest). No strain or species of algae can be considered the best in terms of oil yield for biodiesel. Diatoms and green algae, however, appear to be the most promising (Bajpai 2019). *Scenedesmus dimorphus* is a unicellular alga in the class *Chlorophyceae* (green algae) (Dhaman and Roy 2013; Markou et al. 2013; Sharma and Arya 2017; Show et al. 2017; Leite et al. 2013; Kagan 2010; Chen et al. 2010; Darzins et al. 2010; Patil et al. 2008; Lam and Lee 2012; Jones and Mayfield 2012; Dragone et al. 2010).

Microalgae can provide several different types of renewable biofuels. This includes methane, biodiesel, and biohydrogen (Kapdan and Kargi 2006; Spolaore et al. 2006; Gavrilescu and Chisti 2005). There are many advantages of producing biofuel from algae (Table 1.3). Microalgae can produce 15–300 times more biodiesel than traditional crops on an area basis (Dragone et al., 2010). The harvesting cycle of microalgae is very short and the growth rate is very high (Dragone et al., 2010; Schenk et al. 2008). Moreover, high-quality agricultural land is not required for microalgae biomass production (Scott et al. 2010).

According to Cornell University (Carvalho et al. 2006), fourth-generation biofuels are made using specially created plants or biomass that have either smaller barriers to cellulosic breakdown or greater yields. Additionally, they can be developed on land and water bodies that are unfit for agriculture; thus, no destruction of biomass is warranted (Benemann et al. 1987). It is generated using petroleum-like hydroprocessing (Eriksen et al. 1998). In order for an alternative fuel to be considered a suitable substitute for fossil fuel, it should possess greater environmental benefits over its displaced former, be cost-competitive, and be producible in sufficient amounts to have a meaningful impact on energy demands (Tredici 1999). Most importantly, the net energy derived from the feedstock should exceed the amount that is required for production.

In recent years, scientists have made a breakthrough in this field by designing eucalyptus trees that are able to accumulate three times more carbon dioxide than usual, increasing optimism in mankind's bid to reduce greenhouse gases and salvage the current state of global warming (Biopact 2007). Next, microbes, or microorganisms, are deemed as great alternatives to conventional feedstock for biofuels due to their short life cycle, lower labor requirements, reduced influence by location, season, and climate, as well as the ease of scaling up production (Li et al. 2008).

Fourth-generation biofuels are envisioned as sustainable fuels that obtain higher energy efficiency and environmental performance. This group includes biofuels which can be made using non-arable land. These do not need destroying of biomass to be converted to fuel. The aim of this technology is to directly convert available solar energy to fuel using inexhaustible, inexpensive, and widely available resources. Photobiological solar fuels and electrofuels are the most advanced biofuels which are presently being researched (Aro 2016). According to Aro (2016), "the fourth-generation biofuels photobiological solar fuels and electrofuels—are expected to bring major breakthroughs in the field of biofuels. Technology for the production of such solar biofuels is an emerging area and is based on the direct conversion of solar energy into fuel using raw materials that are inexhaustible, cheap, and widely available. This is expected to take place via the revolutionary development of synthetic biology as an enabling technology for such a change. The synthetic biology field is still in its beginning and only a few truly synthetic examples have been published so far (Cameron et al. 2014). For successful progress, there is a need to discover new-to-nature solutions and construct synthetic living factories and designer microbes for effective and direct conversion of solar energy to fuel. In the same way, a combination of photovoltaics or inorganic water-splitting catalysts with metabolically engineered microbial fuel production pathways (electrobiofuels) is a powerful emerging technology for efficient production and storage of liquid fuels".

Fourth-generation biofuels use raw materials that are unlimited, inexpensive, and available in abundance. Photosynthetic splitting of water into its constituents by solar energy can become a large contributor to fuel production globally, both by artificial photosynthesis and by direct solar biofuel production methods (Inganäs and Sundström 2016). Not only the production of hydrogen but also the production of reduced carbon-based biofuels is possible by concomitant increased fixation of atmospheric carbon dioxide and the innovative design of synthetic metabolic pathways for fuel production. The production of "designer bacteria" with new useful properties needs revolutionary scientific breakthroughs in many areas of fundamental research (link.springer.com).

Fourth-generation biofuels take advantage of the synthetic biology of algae and cyanobacteria. It is a young but strongly evolving research field. Synthetic biology involves the design and construction of new biological parts, devices, and systems and the redesign of existing, natural biological systems for useful applications. It is becoming possible to design a photosynthetic/non-photosynthetic chassis, either natural or synthetic, for producing high-quality biofuels with high PFCE. For the first-, second-, and third-generation biofuels, the raw material is either biomass or waste, both being results of yesterday's photosynthesis. While these biofuels often are very useful in a certain region, they are always limited by the availability of the corresponding biomass, which limits their use worldwide (Aro 2016; Berla et al. 2013; Hays and Ducat 2015; Scaife et al. 2015).

"Fourth-generation biofuels are produced by.

1. designer photosynthetic microorganisms to produce photobiological solar fuels;
2. combining photovoltaics and microbial fuel production (electrobiofuels);
3. synthetic cell factories or synthetic organelles specifically tailored for the production of desired high-value chemicals (production currently based on fossil fuels) and biofuels" (Aro 2016).

The fourth-generation biofuel production process utilizes bioengineering techniques to modify algal metabolism and properties to produce biofuel from photosynthetic organisms (Ale et al. 2019). They are aimed at simultaneously producing bioenergy and capturing and storing carbon dioxide during the production stages (Ben-Iwo et al. 2016). The geosequestered carbon dioxide that is stored in exhausted oil/gas fields, mineral storage (as carbonates), and saline aquifers reduces carbon dioxide emissions and hence makes the fourth-generation biofuel production carbon-negative. These biofuels, therefore, require a minimum number of steps between the sun providing energy and the transformation of this energy into biofuel, thereby avoiding costly fermentation and processing procedures (Lu et al. 2011). This is an emerging field, and extensive research is being undertaken to discover new solutions for the sustainable conversion of energy to fuel.

The difference between the fourth-generation biofuel production compared to the second- and third-generation biofuel production is that the former captures carbon dioxide emissions at all stages of the biofuels production process by means of oxy-fuel combustion (Ziolkowska 2020; Oh et al. 2018; Sher et al. 2018). Oxy-fuel combustion is a process utilizing oxygen (rather than air) for combustion yielding flue gas carbon dioxide and water (Markewitz et al. 2012). While the process is more effective in generating carbon dioxide stream of a higher concentration (the mass and volume are reduced by about 75%), making it more suitable for carbon sequestration, the economic problem occurs mainly at the initial stage of separating oxygen from the air and using it for combustion. The process requires high-energy inputs; nearly 15% of the production of a coal-fired power station can be consumed for this process, which can ultimately drive up production costs and make the final process economically infeasible. Even though currently still not competitive, oxy-fuel combustion has been studied as a potential alternative in combination with biofuels production. For this reason, this technology is in the developing stage as of today. However, if successfully validated in the future, it could be used to geosequester carbon dioxide by storing it in old oil and gas fields or saline aquifers. In this way, through carbon capturing and storage, the fourth-generation biofuels production could be called carbon-negative rather than carbon–neutral. Thus, environmental advantages arise both from carbon storage and from replacing fossil fuels with biofuels. The remaining fuel from oxy-fuel combustion is cleaned and liquefied and yields ultraclean biohydrogen, biomethane, or synthetic biofuels that can be used in the transport sector as well as for electricity generation.

References

D.L. Aguilar, R.M. Rodríguez-Jasso, E. Zanuso, A.A. Lara-Flores, C.N. Aguilar, A. Sanchez, H.A. Ruiz, Operational strategies for enzymatic hydrolysis in a biorefinery, in *Biorefining of Biomass to Biofuels: Opportunities and Perception*. ed. by S. Kumar, R.K. Sani (Springer, Cham, 2018), pp. 223–248

F. Alam, A. Date, R. Rasjidin, S. Mobin, H. Moria, A. Baqui, Biofuel from algae—Is it a viable alternative? Procedia Eng. **49**, 221–227 (2012)

S. Ale, P.V. Femeena, S. Mehan, R. Cibin, Environmental impacts of bioenergy crop production and benefits of multifunctional bioenergy systems. In *Bioenergy with Carbon Capture and Storage* (Elsevier, 2019)

S. Amin, Review on biofuel oil and gas production processes from microalgae. Energy Convers. Manage. **50**, 1834 (2009)

E.M. Aro, From first generation biofuels to advanced solar biofuels. Ambio **45**(Suppl. 1), S24–S31 (2016)

P. Bajpai, *Advances in Bioethanol* (Springer, 2013). ISBN-13 978-8132215837

P. Bajpai, *Third Generation Biofuels* (SpringerBriefs in Energy, Springer Singapore, 2019)

M. Balat, Production of bioethanol from lignocellulosic materials via the biochemical pathway: a review. Energy Convers. Manag. **52**, 858–875 (2011)

J. Ben-Iwo, V. Manovic, P. Longhurst, Biomass resources and biofuels potential for the production of transportation fuels in Nigeria. Renew. Sustain. Energy Rev. **63**, 172–192 (2016)

S. Behera, R.C. Mohanty, R.C. Ray, Batch ethanol production from cassava (Manihotesculenta Crantz.) flourusing Saccharomyces cerevisiae cells immobilized in calcium alginate. Ann Microbiol **65**, 779–783 (2014)

S. Behera, R. Singh, R. Arora, N.K. Sharma, M. Shukla, S. Kumar, Scope of algae as third generation biofuels, Frontiers in bioengineering and biotechnology. Marine Biotechnol. **90**(2), 1–13 (2015)

J.R. Benemann, D.M. Tillett, J.C. Weissman, Microalgae biotechnology. Trends Biotechnol. **5**, 47–53 (1987)

B.M. Berla, R. Saha, C.M. Immethun, C.D. Maranas, T.S. Moon, H.B. Pakrasi, Synthetic biology of cyanobacteria: unique challenges and opportunities. Front. Microbiol. **4**, 246 (2013). https://doi.org/10.3389/fmicb.2013.00246

Biofuels Digest, Mossi and Ghisolf to build 66 Mgy ethanol plant in Piedmont, Italy. 6 February. http://www.biofuelsdigest.com (2008)

Biopact, A quick look at fourth generation biofuels. Retrieved 9 Jan 2014, from Biopact: http://news.mongabay.com/bioenergy/2007/10/quick-look-at-fourth-generation.html (2007)

P. Borse, A. Sheth, Technological and commercial update for first-and second generation ethanol production in India, in *Sustainable Biofuels Development in India*. ed. by A.K. Chandel, R.K. Sukumaran (Springer, 2017), pp. 279–297

A.N. Boyce, P. Chowd-Hury, M. Naqiuddin, Biodiesel fuel production from algae a srenewable energy. Am. J. Biochem. Biotechnol. **4**, 250–254 (2008)

D.E. Cameron, C.J. Bashor, J.J. Collins, A brief history of synthetic biology. Nat. Rev. Microbiol. **12**, 381–390 (2014)

A. Carroll, C. Somerville, Cellulosic biofuels. Annu. Rev. Plant Biol. **60**, 165–182 (2009)

A.P. Carvalho, L.A. Meireles, F.X. Malcata, Microalgal reactors: a review of enclosed system designs and performances. Biotechnol. Prog. **22**, 1490–1506 (2006)

A. Cesaro, V. Belgiorno, Combined biogas and bioethanol production: opportunities and challenges for industrial application. Energies **8**(8), 8121–8144 (2015)

J. Chen, Y. Lu, L. Guo, X. Zhang, P. Xiao, Hydrogen production by biomass gasification in supercritical water using concentrated solar energy: system development and proof of concept. Int. J. Hydrogen Energy **35**, 7134–7141 (2010)

E.C. Clausen, J.L. Gaddy, Ethanol from biomass by gasification/fermentation. http://www.anl.gov/PCS/acsfuel/preprint%20archive/Files/38_3_CHICAGO_08-93_0855.pdf (2009)

A. Darzins, P. Pienkos, L. Edye, *Current Status and Potential for Algal Biofuel Production* (IEA Bioenergy Task, 2010), p. 39

Y. Dhaman, P. Roy, Challenges and generations of biofuels: will algae fuel the world? Ferment Technol. **2**, 119 (2013). https://doi.org/10.4172/2167-7972.1000119

G. Dragone, B. Fernandes, A.A. Vicente, J.A. Teixeira, Third generation biofuels from microalgae, in *Current Research, Technology and Education Topics in Applied Microbiology and Microbial Biotechnology*, ed. by A. Mendez-Vilas (Formatex, Madrid, 2010), pp. 1315–1366

N. Eriksen, B. Poulsen, I.J. Lønsmann, Dual sparging laboratory-scale photobioreactor for continuous production of microalgae. J. Appl. Phycol. **10**, 377–382 (1998)

A.E. Farell, R.J. Plevin, B.T. Turner, A.D. Jones, M. O'Hare, D.M. Kammen, Ethanol can contribute to energy and environmental goals. Science **311**(5760), 506–508 (2006)

M. Gavrilescu, Y. Chisti, Biotechnology—A sustainable alternative for chemical industry. Biotechnol Adv **23**, 471–499 (2005)

E. Gnansounou, Production and use of lignocellulosic bioethanol in Europe: current situation and perspectives. Bioresour. Technol. **101**(13), 4842–4850

L.D. Gomez, C.G. Stelle-King, S.J. McQueen-Mason, Sustainable liquid biofuels from biomass: the writing's on the walls. New Phytol. **178**, 473–485 (2008)

A. Graf, T. Koehler, *Oregon Cellulose-Ethanol Study* (Oregon Office of Energy, Salem, OR, USA, 2000)

S.G. Hays, D.C. Ducat, Engineering cyanobacteria as photosynthetic feedstock factories. Photosynth. Res. **123**, 285–295 (2015)

G.W. Huber, S. Iborra, A. Corma, Synthesis of transportation fuels from biomass: chemistry, catalysts, and engineering. Chem. Rev. **106**, 4044–4098 (2006)

A.D. Hughes, M.S. Kelly, K.D. Black, M.S. Stanley, Biogas from macroalgae: is it time to revisit the idea? Biotechnol. Biofuels **5**, 1–7 (2012)

R.L. Huhnke, Cellulosic ethanol using gasification-fermentation. Resour.: Eng. Technol. Sustain. World: http://www.articlearchives.com/energy-utilities/renewable-energy-biomass/896 186-1 (2009)

O. Inganäs, Sundström, Solar energy for electricity and fuels. Ambio **45**(S1), 15–23 (2016). https://doi.org/10.1007/s13280-015-0729-6

International Energy Agency, From 1st- to 2nd-generation biofuel technologies—An overview of current industry and RDandD activities. (International Energy Agency, IEA/OECD, Paris, 2008), p. 120. www.iea.org

C.S. Jones, S.P. Mayfield, Algae biofuels: versatility for the future of bioenergy. Curr. Opin. Biotechnol. **23**, 346–351 (2012). https://doi.org/10.1016/j.copbio.2011.10.013

J. Kagan, Third and fourth generation biofuels: technologies, markets and economics through 2015 (2010)

I.K. Kapdan, F. Kargi, Bio-hydrogen production from waste materials. Enzyme Microb. Technol. **38**, 569–582 (2006)

M.K. Lam, K.T. Lee, Microalgae biofuels: a critical review of issues, problems and the way forward. Biotechnol. Adv. **30**, 630–690 (2012). https://doi.org/10.1016/j.biotechadv.2011.11.008

R.A. Lee, J.M. Lavoie, From first-to third-generation biofuels: challenges of producing a commodity from a biomass of increasing complexity. Anim. Front. **3**, 6–11 (2013)

G.B. Leite, A.E. Abdelaziz, P.C. Hallenbeck, Algal biofuels: challenges and opportunities. Bioresour. Technol. **145**, 134–141 (2013)

Q. Li, W. Du, D. Liu, Perspective of microbial oils for biodiesel production. Appl. Microbiol. Biotechnol. **80**, 749–756 (2008)

J. Lu, C. Sheahan, P. Fu, Metabolic engineering of algae for fourth generation biofuels production. Energy Environ. Sci. **4**, 2451–2466 (2011)

L.R. Lynd, Overview and evaluation of fuel ethanol from cellulosic biomass: technology, economics, the environment, and policy. Ann. Rev. Energy Environ. **1996**(21), 403–465 (1996)

W.E. Mabee, J. Saddler, Deployment of 2nd-generation biofuels, in *Technology Learning and Deployment Workshop, International Energy Agency, Paris, 11–12 June*. http://www.iea.org/tex tbase/work/2007/learning/agenda.pdf (Gomez LD, Clare GS, McQueen, 2007)

P. Markewitz, W. Leitner, J. Linssen, P. Zapp, T. M€uller, A. Schreiber, Worldwide innovations in the development of carbon capture technologies and the utilization of CO_2. Energy Environ. Sci. **6**, 7281–7385 (2012)

G. Markou, I. Angelidaki, E. Nerantzis, D. Georgakakis, Bioethanol production by carbohydrate-enriched biomass of Arthrospira (Spirulina) platensis. Energies **2013**(6), 3937–3950 (2013)

S.N. Naik, V.V. Goud, P.K Rout, A.K. Dalai, Production of first and second generation biofuels: a comprehensive review. Renew. Sustain. Energy Rev. **14**, 578–597

Y.K. Oh, K.-R. Hwang, C. Kim, J.R. Kim, J.S. Lee, Recent developments and key barriers to advanced biofuels: a short review. Bioresour. Technol. **257**, 320–333 (2018)

R. Olaganathan, F. Ko Qui Shen, L. Jun Shen, Potential and technological advancement of biofuels. Int. J. Adv. Sci. Tech. Res. **4**(4). Retrieved from https://commons.erau.edu/publication/833

V. Patil, K.Q. Tran, H.R. Giselräd, Towards sustainable production of biofuels from microalgae. Int. J. Mol. Sci. **9**(7), 1188–1195 (2008)

Y. Pu, D. Zhang, P.M. Singh, A.J. Ragauskas, The new forestry biofuels sector. Biofuels Bioprod Bioref **2**, 58–73 (2008).

A.J. Ragauskas, M. Nagy, D.H. Kim, C.A. Eckert, J.P. Hallett, C.L. Liotta, From wood to fuels, integrating biofuels and pulp production. Ind. Biotechnol. **2**, 55–65 (2006a)

A.J. Ragauskas, C.K. Williams, B.H. Davison, G. Britovsek, J. Cairney, E.C.A. Frederick, W.J. Jr, J.P. Hallett, D.J. Leak, The path forward for biofuels and biomaterials. Science **311**, 484–489 (2006b)

M.A. Scaife, G.T. Nguyen, J. Rico, D. Lambert, K.E. Helliwell, A.G. Smith, Establishing *Chlamydomonas reinhardtii* as an industrial biotechnology host. Plant J. **82**, 532–546 (2015)

P. Schenk, S. Thomas-Hall, E. Stephens, U. Marx, J. Mussgnug, C. Posten, O. Kruse, B. Hankamer, Second generation biofuels: high efficiency microalgae for biodiesel production. BioEnergy Res. **1**, 20–43 (2008)

S.A. Scott, M.P. Davey, J.S. Dennis, I. Horst, C.J. Howe, D.J. Lea-Smith, A.G. Smith, Biodiesel from algae: challenges and prospects. Curr. Opin. Biotechnol. **21**, 277–286 (2010)

S. Shafiee, E. Topal. When will fossil fuel reserves be diminished? Energy Policy **37**(1), 181–189 (2009)

A. Sharma, S.K. Arya, Hydrogen from algal biomass: a review of production process. Biotechnol. Rep. **14**, 63–69 (2017)

F. Sher, M.A. Pans, C. Sun, C. Snape, H. Liu, Oxy-fuel combustion study of biomass fuels in a 20 kWth fluidized bed combustor. Fuel **215**, 778–786 (2018)

P.L. Show, M.S.Y. Tang, D. Nagarajan, T.C. Ling, C.W. Ooi, J.S. Chang, A holistic approach to managing microalgae for biofuel applications. Int. J. Mol. Sci. **18**, 215 (2017). https://doi.org/10.3390/ijms18010215

Z. Shuping, W. Yulong, Y. Mingde, I. Kaleem, L. Chun, J. Tong, Production and characterization of bio-oil from hydrothermal liquefaction of micro algae Dunaliella tertiolecta cake. Energy **35**, 5406–5411 (2010)

A. Singh, S.I. Olsen, P.S. Nigam, A viable technology to generate third generation biofuel. J. Chem. Technol. Biotechnol. **86**, 1349–1353 (2011)

R. Singh, S. Behera, Y.K. Yadav, S. Kumar, Potential of wheat straw for biogas production using thermophiles. *Recent Advances in Bio-energy Research*. ed. by S. Kumar, A.K. Sarma, S.K. Tyagi, Y.K. Yadav (SSS—National Institute of Renewable Energy, Kapurthala, 2014), pp. 242–249

G. Sivakumar, D.R. Vail, J.F. Xu, D.M. Burner, J.O. Lay Jr., X.M. Ge, P.J. Weathers, Bioethanol and biodiesel: alternative liquid fuels for future generations. Eng. Life Sci. **10**(1), 8–18 (2010)

P. Spolaore, C. Joannis-Cassan, E. Duran, A. Isambert, Commercial applications of microalgae. J. Biosci. Bioeng. **101**, 87–96 (2006)

M.R. Tredici, Bioreactors, photo, in *Encyclopedia of Bioprocess Technology: Fermentation, Biocatalysis, and Bioseparation*. ed. by M.C. Flickinger, S.W. Drew (Wiley, New York, NY, 1999), pp. 395–419

K. Ullah, V.K. Sharma, M. Ahmad, P. Lv, J. Krahl, Z. Wang, Sofia, The insight views of advanced technologies and its application in bio-origin fuel synthesis from lignocelluloses biomasses waste, a review. Renew. Sustain. Energy Rev. **82**, pp. 3992–4008 (2018)

United Nations Environmental Programme, *Towards Sustainable Production and Use of Resources: Assessing Biofuels*. (International Panel for Sustainable Resource Management, France). Retrieved 4 Feb 2014 (2009)

G.M. Walker, Bioethanol: science and technology of fuel alcohol (© 2010 Graeme M. Walker and Ventus Publishing ApS, 2010) ISBN 978-87-7681-681-0

R. Wooley, M. Ruth, J. Sheehan, K. Ibsen, H. Majdeski, A. Galvez, Lignocellulosic biomass to ethanol process design and economics utilizing co-current dilute acid prehydrolysis and enzymatic hyrolysis—Current and futuristic scenarios, Report No. TP-580-26157 (National Renewable Energy Laboratory, Golden Colorade USA, 1999), 130 pp

L. Yanqun, H. Mark, W. Nan, Q.L. Christopher, D.C. Nathalie, Biofuels from microalgae. Biotechnol. Prog. **24**, 815–820 (2008)

J.R. Ziolkowska, Introduction to biofuels and potentials of nanotechnology, in *Green Nanotechnology for Biofuel Production. Biofuel and Biorefinery Technologies*. ed. by N. Srivastava, M. Srivastava, H. Pandey, P.K. Mishra, P.W. Ramteke (Springer, Basel, 2018), pp. 1–15

J.R. Ziolkowska, Biofuels technologies: an overview of feedstocks, processes, and technologies, in *Book: Biofuels for a More Sustainable Future* (Elsevier, 2020)

Chapter 2
Genetically Modified (GM) Microalgae for Biofuel Production

Generally, algae are referred to as plant-like organisms which are usually photosynthetic and aquatic. These have simple reproductive structures and do not have stems, true roots, leaves, and vascular tissue. They are found in the sea, in freshwater, and in moist situations on land. Most of the algae are microscopic, but a few are quite large, for example, some marine seaweeds which may exceed 50 m in length. The algae contain chlorophyll and are able to produce their own food by the process of photosynthesis.

Different groups of algae are presented in Table 2.1. Different options are available for algae type and strain selection.

"No feedstock matches algae when it comes to the potential to produce fuel", (biofuel.org.uk). The algal oil can be used to produce diesel or gasoline and genetic modification can also be performed (slideplayer.com).

Microalgae are one of the most promising species in alternative feedstock for renewable energy production; they can contribute not only to develop sustainable energy resources, but also protect the environment from air pollution and global warming through their sequestration of carbon dioxide (Zhu et al. 2017). Microalgae include prokaryotic cyanobacteria and eukaryotic algae such as *Chlamydomonas* and *Nannochloropsis*. Microalgae usually have a simple life cycle, high growth rate, easy genetic tractability, impressive photosynthetic capability, fertile land independency, and high area yield of valuable co-products; these advantages render them attractive feedstock of bioenergy.

Bioengineers are forecasting that microalgae will be redesigned for producing biofuels using the insights from synthetic biology which is an advanced method of producing genetically engineered organisms (Dana et al. 2012).

Several algal species are found to be suitable for mass culture (Roy and Pal 2015; Guedes and Malcata 2012), mainly aquaculture related, and also including the production of fine chemicals, biofertiliser, biofuels, food and feeds, pharmaceuticals, and wastewater treatment (Chaumont 1993; Becker 2007; Sutherland et al. 2015; Singh et al. 2017). "The advent of the genomic era has heralded a new dawn in

© The Author(s), under exclusive license to Springer Nature Singapore Pte Ltd. 2022
P. Bajpai, *Fourth Generation Biofuels*,
SpringerBriefs in Applied Sciences and Technology,
https://doi.org/10.1007/978-981-19-2001-1_2

Table 2.1 Different groups of algae

Phaeophyceae (macroalgae) Habitat: marine storage: taminarin Pigments: chlorophyll A&C Cell wall: cellulose
Rhodophyceae (microalgae and macroalgae) Habitat: marine/freshwater Storage: Floridean Pigments: chlorophyll A&D Cell well: cellulose (IN)/mucilaginous (OUT)
Chlorophyceae (microalgae and macroalgae) Habitat: marine/freshwater Storage: starch Pigments: chlorophyll A&B Cell wall: cellulose (IN)/pectose (OUT)
Xanthophyceae (microalgae) Habitat: marine/freshwater Storage: Leucosin Pigments: chlorophyll A Cell wall: cellulose and hemicellulose
Euglenophyceae (microalgae) Habitat: freshwater Storage: pararnylon Pigments: chlorophyll A&B Cell wall: N/A
Pyrrophyceae (microalgae) Habitat: marine/freshwater Storage: starch Pigments: chlorophyll A&C Cell wall: has stiff cellulose plates on the outer surface
Chrysophyceae (microalgae) Habitat: marine/freshwater Storage: Leucosin and oil drops Pigments: chlorophyll A&C Cell wall: cellulose with silicate frustules
Cyanophyceae (microalgae) Habitat: marine/freshwater Storage: starch and protein Pigments: chlorophyll A&D Cell wall: mucopeptide with carbohydrates, amino acids and fatty acids

Saad et al. (2019)

microalgal exploitation potential by allowing the combination and selection of key physiological characteristics with modified metabolic activities, increasing production of native compounds relative to wild-type strains, or introducing genes for producing additional non-native compounds or added functionality. Microalgae have been commercially cultured for well over 40 years and the systems presently used at scale tend to be unsophisticated shallow open ponds with no artificial mixing or, alternatively, paddle wheel mixed raceway ponds, both of which can cover hundreds of hectares in size (Borowitzka 1999). Commercialisation of GM microalgae will inevitably require growing GM microalgae on a large scale, but this will require more rigorous risk assessment and environmental management strategies than those used for the unmodified wild-type algae presently being grown" (Beacham et al. 2017).

The fourth-generation biofuel (FGB) is achieved by upgrading the quality and productivity of microalgae by using genetic modification. A microalgae comprise molecules of lipids, proteins, carbohydrates, and nucleic acids (Demirbas and Demirbas 2011). Lipids contain the highest energy level (37.6 kJ g^{-1}) among biochemical components of microalgae. The lipid content in the *biochemical composition* of *microalgae* can be in any form of polar or neutral lipids. Polar lipids constitute the structural lipid by bounding into the organelle membranes or in the bilayer structure of the cell membrane. Neutral lipids are storage lipids, which may include triglycerides, diglycerides, monoglycerides, and free fatty acids (Sajjadi et al. 2018). So obtaining a higher lipid content from algae plays an important role in the efficiency of FGB production (Nwokoagbara et al. 2015).

Three different strategies are generally used for the overproduction of lipids in microalgal strains (Shokravi et al. 2019). These are biochemical, genetic, and transcription methods (Courchesne et al. 2009). The efficiency of the selected strategy is very much dependent on the suitability of the microalgae species for the purpose (Nwokoagbara et al. 2015).

1. Biochemical strategies use environmental stresses during the growing process and mostly deal with controlling the nutrient, salinity, mineral, chemical, and physical factors of microalgae cultivation (Chen et al. 2017).
2. Genetic engineering of microalgae enables further improvement of production efficiency. Understanding the metabolism pathway is considered to be a basis for fully using the potential of microalgae as a lipid source. Genetic engineering involves several strategies in FGB for genome editing, like gene knockdown, knockout, and genetic modifications (Banerjee et al. 2018).
3. Transcription factor engineering deals with the regulation of metabolic pathways within whole cells rather than at the single pathway level. It is conducted by controlling the abundance or function of various enzymes relevant to the production of the desired constituent in microalgae (Capell and Christou 2004).

2.1 Selection of Microalgal Strain

The selection of a suitable microalgal strain is an important step in the production of biofuel. The importance of strain selection is based on the requirement to meet higher biomass productivity, higher content of energetic compounds in one single organism, and associated process characteristics like a feasible medium for growth, lower requirement of nutrients, and facilitated harvesting. An increase in the isolation and screening of microalgae for biofuel production has been seen after 2013 (Sydney et al. 2019). The biomass production in different photobioreactors is limited not only by strain type, but by external conditions such as self-shadowing and photosynthetic productivity. The selection of new strains for producing biofuel could focus on the capacity to store large amounts of desired storage compounds and specific growth conditions like stress conditions, harvesting, temperature, salt tolerance, and straightforward genetic manipulation (Vandamme et al. 2013).

"Biofuels are low value-added products and this is the key criterion for the development of processes and technologies for their production. Intracellular lipid/carbohydratecontent is one of the characteristics that plays a central role in the economical and technical viability of microalgae biofuel processes, in addition to biomass productivity and harvesting. Biomass productivity and improved lipid/carbohydratecontent within cells are directly linked to the production capacity, while new alternative harvesting technologies, or the selection of strains capable of autoflocculating, are crucial due to the high cost of traditional methods (centrifugation and filtration) caused by the small size of the cells. Many strategies and engineering processes have been developed in recent decades, but none has been sufficient to achieve economic viability. Appropriate screening, selection, and genetic manipulation of strains are gaining increasing importance as key aspects in technological development and will probably be crucial for industrial production of microalgae biofuels" (Sydney et al. 2019).

According to Shokravi et al. (2019), the following parameters have an important effect on the appropriateness of microalgae for producing biofuel:

- Growth rate.
- Growth condition.
- Chemical composition.
- Digestibility.

Lipid is the major constituent in microalgae which is used for producing biodiesel. So far biodiesel is the only alternative fuel that can be used directly in the combustion of existing engines. From the transesterification of the lipids present in microalgal cells, biodiesel can be produced (Robles-Medina et al. 2009). Biodiesel is found to be well-suited to conventional diesel fuel. It can be used as an alternative or as an admixture blended in any amount (Lapuerta et al. 2017). Biodiesel is found to reduce the amount of soot in consumption but it does not offer any improvement in ignition quality. The low soot production in biodiesel fuel is due to the presence of ester moiety in biodiesel and the absence of aromatic species (Li et al. 2018). The carbohydrate fraction of microalgae mainly gets originated from the starch present in the plastids and cellulosic polysaccharides in the cell wall (Ravindran et al. 2017). The absence of lignin and the presence of low content of hemicelluloses makes carbohydrates of microalgae a better choice for producing biofuel (Chen et al. 2013).

The composition and the metabolism of carbohydrates vary considerably from species to species. Because of this reason, great care should be taken during the selection of suitable strain for producing biofuel. The higher productivity and the composition of carbohydrates are the most important parameters for increasing the efficiency of biofuel production (Shokravi et al. 2019).

Some strains of algae like *Scenedesmus, Dunaliella, Chlorella, and Chlamydomonas* are able to accumulate more than 50% starch. Starch and cellulose are found to be the most suitable raw material for producing bioethanol and biobutanol. The energy content of carbohydrates is 15.7 kJ g^{-1} and that of protein is 16.7 kJ g^{-1} compounds which are lower as compared to lipids (Shokravi et al. 2019). For producing biofuels, proteins are generally not used as it is difficult to deaminate protein hydrolysates (Chen et al. 2017). The properties of some of the most widely used microalgae for biofuel production are presented in Table 2.2.

Genetic modification of microalgae strains can be used for increasing the amount of lipid, starch, and hydrocarbons released by the algae (Shokravi et al. 2019; Radakovits et al. 2010). Despite the fact that genetic engineering is commonly used for increasing the productivity and lipid accumulation of microalgal strains, these methods suffer from some drawbacks. Genetic modification cannot be used for all species due to the following reasons (Greenwell et al. 2010):

- Lack of available genomic data.
- Complexity of transgenesis.
- Complications in establishing the delicate balance between metabolic and energy storage pathways.

Table 2.2 Properties of some of the microalgal strains used for biofuel production

Microalgal strain	Carbohydrates (%)	Lipids (%)	Proteins (%)
Phaeodactylum tricornutum	8.4	18–57	30
Spirulina platensis	8–14	4–9	46–63
Scenedesmus quadricauda	12.0	1.9	40–47
Chlorella protothecoides	11–15	40–60	10–28
Chlorella minutissima	8.06	14–57	47.89
Scenedesmus obliquus	20–40	30–50	10–45
Scenedesmus dimorphus	21–52	16–40	8–18
Tetraselmis suecica	–	15–23	–
Haematococcus pluvialis	–	25	–
Thalassiosira pseudonana	–	20	–
Dunaliella tertiolecta	12.2–14	11–16	20–29
Dunaliella salina	32	6–25	57
Botlyococcus braunii	–	25	–
Chlorella sorokiniana	26.8	22–24	40.5
Chlorella pyrenoidosa	26	2	57
Chlorella vulgaris	12–17	41–58	12–17

Based on Shokravi et al. (2019), Sajjadi et al. (2018), Greenwell et al. (2010)

Remmers et al. (2018) have reported that almost 20 whole-algal genomes are available which allow for the modification of the targeted genomes for increasing the lipid metabolism.

Microalgae can be used in the form of whole cells, or processed for the extraction of specific components particularly lipids and carbohydrates which are further converted into biofuels. Depending on the desired final product, specific microalgae strains may be selected, screened, or genetically engineered/modified.

The behavior of selected strains under biochemical, genetic, and transcription strategies is of great importance. For example, the use of biochemical stress in the cultivation stage may increase the lipid content, but it simultaneously reduces the growth rate of the microalgal strain (Shokravi et al. 2019).

Cyanobacteria are of particular interest in the production of different types of biofuels, such as bioethanol, biodiesel, biogas, and biohydrogen. The reasons are listed as follows:

- Simple genetic structure.
- Low nutrient demand.
- Broad environmental tolerance.

2.2   Genetic Modification of Microalgae

It is important to describe what "Genetic Modification" means. The term genetically modified organism (GMO) is used to refer to any microorganism, plant, or animal in which genetic engineering techniques have been used to introduce, remove, or modify specific parts of its genome. The techniques which replicate naturally occurring phenomena such as random mutagenesis are not generally considered to result in GMOs under European guidelines and so are not subjected to GM control measures or legislation (Beacham et al. 2017).

Several studies are presently directed toward genetic modification of microalgae for biofuel application as many difficulties are faced in developing economically viable, industrial-scale production of microalgae biofuels using natural strains. "Genetic manipulation of microalgae can lead to major changes in microalgal biotechnology-based industries (Doron et al. 2016). Unfortunately, few microalgae are known to be suitable for genetic manipulation. The green eukaryotic microalgae *Chlamydomonas renhardtii* are largely used as a model organism and the various aspects of metabolic pathway analysis of *C. reinhardtii* have elucidated some novel metabolically important genes (Banerjee et al. 2016). The advances achieved with this microalga are expected to be successfully applied in other strains soon. Recent advances in the development of new, powerful tools allow much more accurate and easier gene modification. CRISPR-Cas9 (Clustered Regularly Interspaced Short Palindromic Repeats associated protein 9), Transcription Activator-Like (TAL), Effector Nucleases (TALEN), and Zinc-Finger Nucleases (ZFN) are techniques being largely applied in genetic engineering of various types of organisms, including microalgae" (Sydney et al. 2019; Chew et al. 2017).

Another approach for improving biofuel production by microalgae is activating or silencing specific genes. These are frequently used to block competing metabolic pathways to increase lipid accumulation (Baek et al. 2016) or biohydrogen evolution (hydrogenase activity, decreasing its oxygen sensitivity, anaerobiosis induction, avoiding competition for electrons from other pathways, and increasing the sources of electrons) (Gimpel et al. 2015).

Microalgae genetic manipulation for increased biofuel production may focus on the following:

(i)   Improving light utilization, resulting in increased growth and productivity—regulation of antenna size, chlorophyll content (Dubini and Ghirardi 2015), and interchangeable conversion between chlorophyll types (Gomaa et al. 2016).

(ii)   Improving carbon dioxide fixation, because lipid and carbohydrate accumulation occurs due to carbon fixation during depletion of essential nutrients for cell division—target enzymes include RuBisCO, sedoheptulose-1,7-bisphosphatase (SBPase), and phosphoenolpyruvate carboxylase(PEPCase) (Gomaa et al. 2016).

(iii)   Improving final product excretion, avoiding high costs for cell lysis—incorporating an inducible lytic system (Miyake et al. 2014).

(iv)   Improving harvesting, reducing biomass recovery costs—overexpression of Extracellular Polymeric Substances (EPS) such as glycoproteins, functional proteins, and polysaccharides (Khademi et al. 2015).

Most recent patents describing genetic modification are aimed at improved biomass productivity and intracellular product content. A very few focus on harvesting methods. The Espacenet (http://www.epo.org) was searched for flocculation and microalgae and only 2 among 84 deposited patents dealt with genetic modification/transformation of microalgae facilitating cell harvesting" (Sydney et al. 2019; Fengwu and Xinqing 2013; Kwa et al. 2013).

Genetic modification of blue-green and green algae has been reported. The green algae *C. reinhardtii* has been genetically engineered to express many important features of biofuels but reduced biomass production rates and reduced lipid content in most of the model strains have prevented them from becoming commercially important (Ahmad et al. 2014). Major green algal species, such as *Scenedesmus, Nannochloropsis, Botryococcus, Parachlorella Chlorella,* and *Neochloris,* are found to be rich in lipid content. So these are potential candidates for biofuel production. Up to now, genetic modification has largely been restricted to formative studies which involve transferring of the reporter genes and/or selectable marker genes, only two species, *Dunaliella* and *Chlorella* transformed with biofuel-related traits (Table 2.3).

Cyanobacteria which is a blue-green algae accumulate relatively lower lipid content in comparison to the green algae. Therefore, these are not important candidates for biofuel production. However, in spite of this, cyanobacteria may be engineered for producing other important biofuel-related molecules because they are able to grow in extreme environments and they can be genetically modified relatively easily. In the case of cyanobacteria, most genetic engineering has been

Table 2.3 Important green algal species which have the potential for biofuel production, and the status of their genetic modification with biofuel-related traits

Algal Species	Genetic transformation with biofuel-related traits	Reference
Chlorella	Hydrogenase (*HydA*)	Chien et al. (2012)
Dunaliella	Xylanase, α- galactosidase, phytase, phosphate anhydrolase, and β-mannanase	Georgianna et al. (2013)
Nannochloropsis	NIL	–
Scenedesmus	NIL	–
Botryococcus	NIL	–
Neochloris	NIL	–

Reproduced with permission from Kumar (2015)

conducted on model organisms, *Synechococcus* sp. PCC 7002, *Synechococcus elongatus* sp. PCC 7492, *Synechocystis* sp. PCC 6803, and *Anabaena* sp. PCC 7120. But, the perfect production host remains difficult to predict (Savakis and Hellingwerf 2015). Cyanobacteria have also been engineered to produce a variety of different biofuel-related molecules (Table 2.4).

Cyanobacteria are able to fix carbon dioxide in a more efficient way as compared to plants and can be engineered for producing carbon feedstocks useful for biofuel. But, the extension of this technology to the industrial scale is not encouraging because of the current low yields (Oliver et al. 2014).

2.3   Genetic Instability in Genetically Modified Algae

"Generally, GM algae should show a high degree of stability (Savakis and Hellingwerf, 2015) but achieving transgene stability in GM algae is a huge challenge, as shown in studies of green and blue-green algal model systems. The storage of *Chlamydomonas* cultures in a solid medium generally results in the loss of interested traits or spontaneous mutations (Hannon et al. 2010; Ahmad et al. 2014). This may be attributed to the inherent genetic instability and high mutation rate of naturally occurring green algal cells. It has been estimated that there are > 24,000 changes (including single nucleotide variations and insertions/deletions) between two common laboratory strains of green algae (Dutcher et al. 2012). This high rate of mutation is largely due to the haploid genome in *Chlamydomonas* (Siaut et al. 2011). It is a major obstacle to generating useful transgenic algal strains due to rapid gene silencing (Cerutti et al. 1997; Fuhrmann et al. 1999; Neupert et al. 2009). In the unicellular green alga *C. reinhardtii*, epigenetic silencing of transgenes occurs at both the transcriptional and post-transcriptional levels. In the case of single-copy transgenes, transcriptional silencing takes place without detectable cytosine methylation of the introduced DNA (Cerutti et al. 1997; Jeong et al. 2002). The introduced

Table 2.4 Blue-green algae genetically modified with biofuel-related traits

Algal species*	Genetic transformation with biofuel-related traits	Reference
Synechocystis sp. PCC6803	Acyl-acyl carrier protein reductase and aldehyde-deformylating oxygenase (alkanes)	Wang et al. (2013)
	Acyl-acyl carrier protein thioesterase gene	Liu et al. (2011)
	Kivd and adhA from *Lactococcus lactis* (for isobutanol)	Varman et al. (2013)
Synechococcus elongatus PCC 7942	Knockout of the FFA-recycling acyl-ACP synthetase and expression of a thioesterase (fatty acids)	Ruffing and Jones (2012)
	Acyl-ACP thioesterase and acetyl-CoA carboxylase (fatty acids)	Ruffing (2013)
	2-methyl-1-butanol (alcohol)	Shen and Liao (2012)
	Acetyl-CoA acetyl transferase (encoded by *thl*), acetoacetyl-CoA transferase (encoded by *atoAD*), acetoacetate decarboxylase (encoded by *adc*), and secondary alcohol dehydrogenase (encoded by *adh*) (for isopropanol)	Kusakabe et al. (2013)
	Acyl-ACP reductase (*Aar*) overexpression (fatty aldehydes)	Kaiser et al. (2013)

* Model cyanobacterial strains
Reproduced with permission from Kumar (2015)

genes are often unstably expressed, leading to the loss or reduced manifestation of newly acquired traits. *Chlamydomonas* cultures are, therefore, most commonly maintained as vegetative cells on an agar-containing medium for short periods of time. To maintain their quality over the long-term, *Chlamydomonas* cultures can be cryopreserved under liquid nitrogen (González-Ballester et al. 2005, 2011).

Cyanobacteria are believed to be the most primitive organisms on Earth and were one of the first organisms to be genetically transformed—*Synechocystis* was successfully transformed in the early 1970s (Shestakov and Khyen 1970). Banack et al. (2012) reported that cyanobacteria naturally produce N-(2-aminoethyl) glycine (AEG), a small molecule, which may play an important role in gene silencing. After the successful creation of the GM *Synechocystis* strain, it was sequentially transformed with a plasmid bearing *pdc/adh* genes. Several isolates containing the complete ethanol production cassette did not, however, demonstrate stable ethanol production (Dexter and Fu 2009), indicating the instability of transgenes in cyanobacteria. Concerns have also been raised about the segregation of transgenes in polyploid strains" (Kumar 2015; Berla et al. 2013).

2.4 Cultivation of Genetically Modified Microalgae

Microalgae are photosynthetic microorganisms that need light and carbon dioxide as energy and carbon sources, respectively. Algae can be autotrophic, heterotrophic, mixotrophic, and photoheterotrophic according to their metabolic pathways (Daliry et al. 2017).

The autotrophic pathway or photosynthesis involves the conversion of inorganic carbon into organic energy in the presence of light (Chen et al. 1994). Autotrophic cultivation of microalgae is a sunlight-driven system that converts carbon dioxide into lipids and other valuable components. Autotrophic cultivation is not an appropriate choice for genetically modified microalgae because of the possible threat to the environment and health of leaving species (Ghosh et al. 2016). The limitation of the autotrophic method can be overcome by using heterotrophic cultivation. Heterotrophic cultivation takes advantage of using organic carbon sources such as glucose and acetate (Tamayo-Ordóñez et al. 2017; Huppe and Turpin 1994). In a mixotrophic pathway, cells can grow autotrophic or heterotrophic based on the available food sources (Andrews 1968). Mixotrophic cultivation is a preferable microalgae growth mode for biomass production (Zhan et al. 2017). In comparison to photoautotrophic and heterotrophic metabolism, mixotrophic cultures have shown several benefits, such as less risk of contamination, reduced cost, and high biomass productivity. Even prone to contaminations, the use of photobioreactors reduces this risk, but increases the cost of the process, which can be offset by the high biomass yield which can reach 5–15 g/L, which is 3–30 times higher as compared to those produced under autotrophic growth conditions (Ngo-mati et al. 2015; Pereira et al. 2019). The photoheterotrophic pathway takes place in the presence of light and organic carbon. Heterotrophic metabolism results in a higher growth rate in comparison to autotrophic metabolism. Mixotrophic metabolism appears to be the best way to get maximum biomass and lipid productivities. In the mixotrophic method, microalgae can drive the autotrophic mode by receiving inorganic carbon from photosynthesis, as well as heterotrophic form by using organic carbon sources such as glucose, glycerol, acetate, maltose, sucrose, and fructose (Chen et al. 1994; Scarsella et al. 2010; Martinez et al. 1991; www.mdpi.com).

For the production of algae on small scale, algae are generally grown in closed systems or photobioreactors. In closed systems, contact between the enclosed algae and the environment is prevented. Genetically modified cyanobacteria can be grown in photobioreactors, which are found to be safe for growing bacterial cultures. Closed systems can also be placed in greenhouses or outdoors to use natural sunlight (Fig. 2.1).

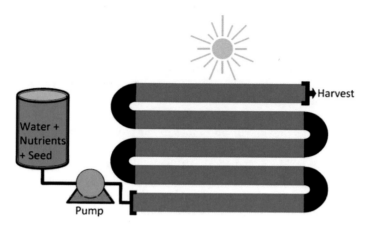

Fig. 2.1 Outdoor cultivation of GM algae in an enclosed system to tap solar energy for growth. Reproduced with permission from Kumar (2015)

A detailed study on the economic potential of using the heterotrophic cultivation of microalgae was performed by Tabernero et al. (2012). This method uses photobioreactors and consumes a large amount of organic compounds. So, it is not economically viable for producing biofuels.

For using genetic modification in heterotrophic cultivation, the strain modification should target up or downstream metabolism (Remmers et al. 2018). The mixotrophic cultivation offers an exceptional opportunity to combine the benefits of autotrophic and heterotrophic cultivation for dealing with the above issues (Lowrey et al. 2015).

The effect of macronutrients and micronutrients on the growth, biomass, and lipid productivities *of Dunaliella tertiolecta* was studied by Mata et al. (2013). With the exception of nitrogen and iron, the other nutrients added to the culture medium such as magnesium, potassium, manganese, and zinc did not affect the studied parameters. The best maximum and average dry biomass productivities were 141.8 mg/L/d and 63.1 mg/L/d, respectively, obtained on the 5th day, by increasing the nitrogen concentration 10 times in comparison to the standard culture medium—artificial seawater medium with vitamins. In these conditions, the lipid content and maximum lipid productivity were 33.5% and 47.4 mg/L/d, respectively. This corresponded to more than three times the value in the standard conditions (13.3 mg/L/d). Also, by increasing the iron concentration by 10 times in the culture medium, the maximum lipid productivity increased to almost the double, i.e., from 14.6 to 28.0 mg/L/d, obtained on the 28th day. Addition of nitrogen and iron to the culture medium resulted in a substantial increase in lipid productivity. This suggests that residual wastewaters rich in nitrogen or iron could be used for growing *Dunaliella tertiolecta* for lipid production.

Four green microalgae (TRG, KB, SK, and PSU) identified as *Botryococcus* spp. were isolated from lakes and freshwater ponds in southern Thailand by Yeesang and Cheirsilp (2011). "In a nitrogen-rich medium, the strains achieved a lipid content of 25.8%, 17.8%, 15.8%, and 5.7%, respectively. A combination of nitrogen deficiency, moderately high light intensity (82.5 μE m^{-2} s^{-1}), and high level of iron (0.74 mM) improved lipid accumulation in TRG, KB, SK, and PSU strains up to 35.9%, 30.2%, 28.4%, and 14.7%, respectively. The lipid contents and plant oil-like fatty acid composition of the microalgae suggested their potential as biodiesel feedstock" (Yeesang and Cheirsilp 2011).

The lipid content in heterotrophic cells of *Chlorella saccharophila* increased by about three times in comparison to that of autotrophic cells. The principal fatty acids in heterotrophic *C. saccharophila* were oleic acid (C18:1) and linoleic acid (C18:2) constituting 34.4% and 30.1% of the total fatty acid contents, respectively (Isleten-Hosoglu et al. 2012).

Biomass productivity of *Dunaliella salina, Nannochloropsis oculata, Spirulina platensis, Chlorella sorokiniana,* and *Scenedesmus obliquus* in mixotrophic culture were 2.2, 1.4, 3.8, 2.4–4.2, and 8.7 times that under photoautotrophic condition (Wang et al. 2014).

The photosynthetic products in microalgae are accumulated in the form of cell structural or storage components. The product accumulation in microalgae can be optimized by nutrient deficiency. Nitrogen and phosphorus are the two major macronutrients for the growth and reproduction of algae. These are normally used in relatively large amounts.

The effect of nitrogen and phosphorous limitation on the production of microalgae has been studied by several researchers (Chen et al. 2017; Kamalanathan et al. 2017; Arora et al. 2016). The productivity and lipid accumulation under nutrient depletion vary based on the strain type and culture condition (Tamayo-Ordóñez et al. 2017).

Ten different strains of *Chlorella* and *Parachlorella* under lipid induction growth conditions in autotrophic laboratory cultures were examined by Přibyl et al. (2012). Significant differences in biomass, lipid productivity, and also in the final content of lipids were observed. The availability of nitrates and/or phosphates strongly affected growth and lipid accumulation in cells by affecting cell division. Nutrient limitation significantly increased lipid productivity up to a maximal value of 1.5 g/l/day. The production of lipids through large-scale cultivation of *Chlorella vulgaris* in a thin- layer photobioreactor, even under suboptimal conditions, was shown by these researchers. The maximal lipid productivity was 0.33 g/l/d, biomass density was 5.7 g/l dry weight, and total lipid content was more than 30% dry weight after 8 days of cultivation. *Chlorella vulgaris* lipids contain fatty acids having a relatively high degree of saturation compared with the canola oil offering a possible option for the use of higher plant oils.

The nitrogen and phosphorous depletion in *Chlorella pyrenoidosa* cultivation increase the productivity and lipid accumulation by 50.32% and 34.29% of dry cell weight, respectively (Fan et al. 2014).

The oleaginous microalgae, *D. tertiolecta, N. oculata, Thalassiosira pseudonana, Odontella aurita,, Isochrysis galbana,* and *Chromulina ochromonoides,* were grown at temperatures of 10 and 20 °C and two nutrient regimes (deplete and replete) by Roleda et al. (2013). A strong negative correlation between growth rate and lipid content was seen for all species. Multiple stressors have no additive effect on lipid accumulation. Total oil content (fatty acid methyl esters, FAMEs, pg cell^{-1}) was increased more by nutrient limitation than by temperature stress. In response to nutrient stress, *N. oculata* was found to be the most robust species with an increase in lipid accumulation of up to three- to four-fold compared to the accumulation under nutrient-sufficient conditions. Although stress conditions resulted in reduced fatty acid unsaturation in most taxa because of the higher triacylglycerol production, a high proportion of eicosapentaenoic acid was maintained in *O. aurita*.

Not much information is available on the effect of phosphorous concentration on lipid accumulation (Tamayo-Ordóñez et al., 2017).

Effects of nitrogen and phosphorus concentrations on growth, nutrients uptake, and lipid accumulation of *Scenedesmus* sp. LX1 which is a freshwater microalga were studied by Xin ct al. (2010). *Scenedesmus* sp. LX1s grew in accordance with the Monod model. The nitrogen and phosphorous saturated maximum growth rate was $2.21 \times 10(6)$ cells mL$(^{-1})$d$(^{-1})$, and the half-saturation constants of nitrogen and phosphorous uptake were 12.1 mg L$(^{-1})$ and 0.27 mg L$(^{-1})$, respectively. In the nitrogen/phosphorus ratio of 5:1–12:1, 83–99% nitrogen and 99% phosphorus were removed. In conditions of nitrogen [2.5 mg L$(^{-1})$] or phosphorus [0.1 mg L$(^{-1})$] limitation, *Scenedesmus* sp.. LX1 was found to accumulate lipids as high as 30% and 53%, respectively, of its algal biomass. But the lipid productivity/unit volume of culture was not improved. Further research should be focused on improving both lipid content and lipid productivity.

The nutritional requirements of *D. tertiolecta* were determined by Chen et al. (2011). Deprivation of phosphorous was found to have a lesser effect on lipid accumulation in comparison to nitrogen. Moreover, the starvation of iron or cobalt improved the accumulation of neutral lipids in *D. tertiolecta*. Other nutrient particles in the production of microalgae include macronutrient calcium, sodium, potassium, and magnesium and also microelements such as zinc, boron, cobalt, manganese, iron, and molybdenum (Khan et al. 2018). However, the nitrogen and phosphorous contents in the medium have a much higher impact on lipid accumulation in comparison to the other micronutrients and macronutrients (Chen et al., 2011).

Many microalgal strains do not produce large amounts of lipids during logarithmic growth. For improving the productivity of the structural lipid, environmental stresses such as a lack of nitrogen, high salinities, and a high amount of irradiance are used for slowing down their proliferation and start producing lipids (Hu et al. 2008). Overexpression of the lipid synthesis pathway genes is a solution for slowing down microalgal proliferation. Whereas the overexpression of the genes involved in fatty acid synthesis has had little success, some interesting results have been obtained through the overexpression of genes involved in TAG assembly (Radakovits et al. 2010). Inhibition of lipid catabolism may also result in reducing proliferation and biomass productivity.

In Table 2.5, some strategies for the genetic modification of the lipid metabolism are presented (Shokravi et al. 2019; Du et al. 2018; Choi et al. 2016; Xue et al. 2015; Salas-Montantes et al. 2018; Zulu, et al. 2017).

Light intensity is another important parameter for lipid accumulation and fatty acid composition. The reaction of microalgae to different light intensities can be defined

Table 2.5 Strategies for the genetic modification of the lipid metabolism

Algae	Gene/platform involved	Nutrient condition	Result
Phaeodactylum tricornutum	Overexpression of *Phaeodactylum tricornutum* ME (PtME)	Nitrogen deprivation	Increased neutral lipid content
Synechococcus elongatus PCC 7942	bicA	Nitrogen starvation	Accelerated development of biosolar cell factories
Chlamydomonas reinhardtii	Overexpressing a DNA-binding-with-one-finger (Dof)-type transcription factor	Nutrient deficiency	Increased total lipids an higher proportion of specific fatty acids
Phaeodactylum tricornutum Pt4	Co-expression of a yeast diacylglycerol acyltransferase (ScDGA1) and a plant oleosin (AtOLEO3)	Nitrogen stress	Increase in triacylglycerol levels
Nannochloropsis oceanica CCMP1779	Overexpressing DGTT in nutrient-replete medium	Nutrient-filled medium	Increased production of triacylglycerol

Shokravi et al. (2019), Du et al. (2018), Choi et al. (2016), Xue et al. (2015), Salas-Montantes et al. (2018), Zulu et al. (2017)

by three phases of photolimitation, photosaturation, and photoinhibition that the growth rate in the associated light intensity is increased, independent, and declined, respectively (Shin et al. 2018;. da Silva Ferreira and Sant'Anna 2017). But, the effect of light intensity and light wavelength on metabolic regulation and growth rate of algae is species-dependent and even strain-specific (Shin et al. 2018).

Atta et al. (2013) studied the effects of various intensities of blue light and its photoperiods on the growth and lipid content of *C. vulgaris* by using LED (Light Emitting Diode) in batch culture. *C. vulgaris* was grown for 13 days at three different light intensities [100, 200, and 300 μmol m$^{(-2)}$s$^{(-1)}$]. Effects of three different light and dark regimes (12:12, 16:08, and 24:00 h Light:Dark) were studied for each light intensity at 25 °C temperature. Maximum lipid content (23.5%) was obtained because of high efficiency and deep penetration of 200 μmol m$^{(-2)}$s$^{(-1)}$ of blue light (12:12 L:D) with improved specific growth [1.26 d$^{(-1)}$] within reduced cultivation time of 8 days. White light could produce 20.9% lipid content in 10 days at 16:08 h L:D.

Teo et al. (2014) examined the effect of different light wavelengths on growth and lipid production of *Tetraselmis* sp. and *Nannochloropsis* sp. Microalgae were grown for 14 days under blue, red, red–blue LED, and white fluorescent light. The growth rate of microalgae was analyzed by spectrophotometer and cell counting and oil production under the improved Nile red method. Microalgae showed a better growth curve under blue wavelength. Besides, *Tetraselmis* sp. and *Nannochloropsis* sp.

under blue wavelength showed a higher growth rate and oil production. Gas chromatography analysis also showed that palmitic acid and stearic acid which are necessary components for biodiesel contribute about 49–51% of total FAME from *Nannochloropsis* sp. and 81–83% of total FAME from *Tetraselmis* sp.

Genetic engineering tools are being used for increasing the photosynthetic efficiency to effectively capture light energy (Schenk et al. 2008). Remmers et al. (2018) attempted to genetically modify the accumulation through impairing photosynthetic machinery. Reducing the number of light-harvesting antenna complex pigments or reducing the chlorophyll antenna size is used in some studies for overcoming the light-saturation effect (Medipally et al. 2015).

Genetic modification could reduce the production cost in FGB by 50% or more. Astaxanthin, fucoxanthin, carotenoid, and polyunsaturated fatty acids contents can be increased by 2 times or 3 times by genetic engineering. Use of safer genetic engineering methods such as mutagenesis or self-cloning for the production of industrially valuable algal products may reduce biosafety issues (Sharon-Gojman et al. 2017).

2.5 Harvesting of Genetically Modified Microalgae

Harvesting is the most important phase in the biofuel production process. "Harvesting of algae refers to the concentration of diluted algal broth until a thick paste is obtained. It is one of the major challenges in algae biodiesel initiatives. This step accounts for about 20–30% of the overall production costs in the algae-based biofuel production process. So, the selection of a proper method for harvesting will have an impact on the overall economics of the process. Harvesting of microalgae is difficult as the size of algae is small. The selection of the effective process depends on the properties and size of the algal strain. Several physical and chemical methods can be used for harvesting (Bajpai 2019; Brennan and Owende 2010; Wang et al. 2008; Demirbas 2007; Show et al. 2017; Shah et al. 2017; Behera and Varma 2016; www.formatex. info; Edzwald 1993; Muylaert et al. 2009; Pushparaj et al. 1993; Pahl et al. 2013) as follows:

- Flocculation.
- Centrifugation.
- Filtration.
- Ultrafiltration.
- Air-flotation.
- Autoflotation".

Microalgae harvesting methods are very much dependent upon the following parameters (Singh and Patidar 2018):

- Type of target microalgae.
- Density and cell size of the algae.
- Characteristics of the final product.

- Reusability of the culture medium for the next cultivation cycle.

Presently, several methods are being used for harvesting of microalgae in order to perform the desired solid–liquid separation, which can be divided into four categories (Mata et al. 2010; Mathimani, and Mallick 2018):

(i)   Biological.
(ii)   Chemical.
(iii) Electrical.
(iv)   Mechanical.

The harvesting process of microalgae can be divided into two main stages, including thickening and dewatering (Pahl et al. 2013).

The objective of the thickening stage is to increase the solid concentration of microalgal suspension and reduce the processing volume. The obtained sludge from the thickening process is quite slushy and requires further processing to proceed into the drying stage, which is carried out at the dewatering stage. There are several thickening and dewatering methods for solid–liquid separation (Barros et al. 2015).

One of the major concerns in the harvesting stage of the genetically modified strains relates to the residual water from the harvesting phase. Discharged water from the harvesting process may contain plasmid or chromosomal DNA from the transgene genetically modified algae which might result in horizontal gene transfer (Abdullah et al. 2019). To deal with this problem, several treatment methods have been proposed. These are reviewed in Abdullah et al. (2019) and Beacham et al. (2017).

Aravanis et al. (2010) introduced an end-to-end algal biofuel production process using a genetically engineered *C. reinhardtii* strain. In the production phase of the proposed biofuel known as "Green Crude", the C. *reinhardtii* strain was cultivated in a raceway pond with an area of around 400.000m2. The harvesting and dewatering processes were done by the membrane and disk stack centrifuges, respectively.

Szyjka et al. (2017) evaluated the ecological risk of the open pond cultivation of genetically engineered algae. It was found that outdoor cultivation of GM microalgae did not create any adverse effects on the environment or the surrounding native algae population.

Genetic modification is an attractive option for improving the yield of valuable products in microalgae at a reduced cost. Several methods are available for improvising the lipid content in microalgae. These enhancement strategies can be generally divided into four groups. These include improving photosynthetic efficiencies, biomass productivity, diversity, and the ability to thrive in diverse ecosystems. Among these, improving the photosynthetic efficiencies and biomass productivity are the most researched. Different stress conditions, including the addition or depletion of nutrients, light, and salinity levels, are used for growing the genetically modified strains. But health and environmental risks are the major concerns in the cultivation and harvesting of genetically modified strains (Shokravi et al. 2019).

References

B. Abdullah, S.A.F. Syed Muhammad, Z. Shokravi, S. Ismail, K.A. Kassim, A.N. Mahmood, M.M.A. Aziz, Fourth generation biofuel: A review on risks and mitigation strategies. Renew. Sustain. Energy Rev. **107**, 37–50 (2019)

I. Ahmad, A.K. Sharma, H. Daniell, S. Kumar, Altered lipid composition and enhanced lipid production in microalga by introduction of brassica diacylglycerol acyltransferase 2. Plant Biotechnol. J. Early View (2014). https://doi.org/10.1111/pbi.12278

J.F. Andrews, A mathematical model for the continuous culture of microorganisms utilizing inhibitory substrates. Biotechnol. Bioenergy **10**, 707–723 (1968)

N. Arora, A. Patel, P.A. Pruthi, V. Pruthi, Synergistic dynamics of nitrogen and phosphorous influences lipid productivity in *Chlorella minutissima* for biodiesel production. Biores. Technol. **213**, 79–87 (2016)

A.M. Aravanis, B.L. Goodall, M. Mendez, J.L. Pyle, J.E. Moreno, Methods and systems for biofuel production, US Patent 12763068 (2010)

M. Atta, A. Idris, A. Bukhari, S. Wahidin, Intensity of blue LED light: a potential stimulus for biomass and lipid content in fresh water microalgae *Chlorella vulgaris*. Biores. Technol. **148**, 373–378 (2013)

K. Baek, D. Kim, J. Jeong, S.J. Sim, A. Melis, J.S. Kim, E. Jin, S. Bae, DNA-free two-gene knockout in *Chlamydomonas reinhardtii* via CRISPR-Cas9 ribonucleoproteins. Sci. Rep. **6**(1), 30620 (2016)

P. Bajpai, *Third Generation Biofuels* (SpringerBriefs in Energy, Springer Singapore, 2019)

S.A. Banack, J.S. Metcalf, L. Jiang, D. Craighead, L.L. Ilag, P.A. Cox, Cyanobacteria produce N-(2-aminoethyl)glycine, a backbone for peptide nucleic acids which may have been the first genetic molecules for life on earth. PLoS ONE **7**(11), e49043 (2012)

A. Banerjee, C. Banerjee, S. Negi, J.S. Chang, P. Shukla, Improvements in algal lipid production: A systems biology and gene editing approach. Crit. Rev. Biotechnol. **38**, 369–385 (2018)

C. Banerjee, K.K. Dubey, P. Shukla, Metabolic engineering of microalgal based biofuel production: prospects and challenges. Front. Microbiol. **7**, 1–8 (2016)

A.I. Barros, A.L. Gonçalves, M. Simões, J.C.M. Pires, Harvesting techniques applied to microalgae: a review. Renew. Sustain. Energy Rev. **41**, 1489–1500 (2015)

T.A. Beacham, J.B. Sweet, M.J. Allen, Large scale cultivation of genetically modified microalgae: a new era for environmental risk assessment. Algal Res. **25**, 90–100 (2017)

E.W. Becker, Micro-algae as a source of protein. Biotechnol. Adv. **25**, 207–210 (2007)

B.M. Berla, R. Saha, C.M. Immethun, C.D. Maranas, T.S. Moon, H.B. Pakrasi, Synthetic biology of cyanobacteria: unique challenges and opportunities. Front. Microbiol. **4**, 246 (2013). https://doi.org/10.3389/fmicb.2013.00246

M.A. Borowitzka, Commercial production of microalgae: ponds, tanks, tubes and fermenters. J. Biotechnol. **70**, 313–321 (1999)

L. Brennan, P. Owende, Biofuels from microalgae—a review of technologies for production, processing, and extractions of biofuels and co-products. Renew. Sust. Energ. Rev. **14**, 557–577 (2010)

T. Capell, P. Christou, Progress in plant metabolic engineering. Curr. Opin. Biotechnol. **15**, 148–154 (2004)

A.C. Guedes, F.X. Malcata, Nutritional value and uses of microalgae in aquaculture, in *InTech* ed. by Z. Muchlisin (2012), pp. 59–79. ISBN: 978-953-307-974-5

H. Cerutti, A.M. Johnson, N.W. Gillham, J.E. Boynton, Epigenetic silencing of a foreign gene in nuclear transformants of Chlamydomonas. Plant Cell **9**, 925–945 (1997)

D. Chaumont, Biotechnology of algal biomass production: a review of systems for outdoor mass culture. J. Appl. Phycol. **5**, 593–604 (1993)

C.Y. Chen, X.Q. Zhao, H.W. Yen, S.H. Ho, C.L. Cheng, D.J. Lee, F.W. Bai, J.S. Chang, Microalgae-based carbohydrates for biofuel production. Biochem. Eng. J. **78**, 1–10 (2013)

M. Chen, H. Tang, H. Ma, T.C. Holland, K.Y. Ng, S.O. Salley, Effect of nutrients on growth and lipid accumulation in the green algae *Dunaliella tertiolecta*. Biores. Technol. **102**, 1649–1655 (2011)

B. Chen, C. Wan, M.A. Mehmood, J.S. Chang, F. Bai, X. Zhao, Manipulating environmental stresses and stress tolerance of microalgae for enhanced production of lipids and value-added products—a review. Biores. Technol. **244**, 1198–1206 (2017)

K.W. Chew, J.Y. Yap, P.L. Show, N.H. Suan, J.C. Juan, T.C. Ling, D.J. Lee, J.S. Chang, Microalgae biorefinery: high value products perspectives. Bioresour. Technol. **229**, 53–62 (2017)

L.F. Chien, T.T. Kuo, B.H. Liu, H.D. Lin, T.Y. Feng, C.C. Huang, Solarto-bioH2 production enhanced by homologous overexpression of hydrogenase in green alga Chlorella sp. DT. Int. J. Hydrogen Energy **37**, 17738–17748 (2012)

S. Choi, B. Park, I.G. Choi, S.J. Sim, S.M. Lee, Y. Um, H.M. Woo, Transcriptome landscape of *Synechococcus elongatus* PCC 7942 for nitrogen starvation responses using RNA-seq. Sci. Rep. **6**, 30584 (2016)

N.M.D. Courchesne, A. Parisien, B. Wang, C.Q. Lan, Enhancement of lipid production using biochemical, genetic and transcription factor engineering approaches. J. Biotechnol. **141**, 31–41 (2009)

V. da Silva Ferreira V, C. Sant'Anna, Impact of culture conditions on the chlorophyll content of microalgae for biotechnological applications. World J. Microbiol. Biotechnol. **33**, 20 (2017)

S. Daliry, A. Hallajsani, J. Mohammadi Roshandeh, H. Nouri, A. Golzary, Investigation of optimal condition for *Chlorella vulgaris* microalgae growth. Glob. J. Environ. Sci. Manag. **3**, 217–230 (2017)

A. Demirbas, Progress and recent trends in biofuels. Prog. Energy Combust. Sci. **33**, 1–18 (2007)

A. Demirbas, M.F. Demirbas, Importance of algae oil as a source of biodiesel. Energy Convers. Manage. **52**, 163–170 (2011)

J. Dexter, P. Fu, Metabolic engineering of cyanobacteria for ethanol production. Energy Environ. Sci. **2**, 857–864 (2009)

L. Doron, N. Segal, M. Shapira, Transgene expression in microalgae—from tools to applications. Front. Plant Sci. **7**, 505 (2016)

Z.Y. Du, J. Alvaro, B. Hyden, K. Zienkiewicz, N. Benning, A. Zienkiewicz, G. Bonito, C. Benning, Enhancing oil production and harvest by combining the marine alga *Nannochloropsis oceanica* and the oleaginous fungus *Mortierella elongata*. Biotechnol. Biofuels **11**, 174 (2018)

A. Dubini, M.L. Ghirardi, Engineering photosynthetic organisms for the production of biohydrogen. Photosynth. Res. **123**(3), 241–253 (2015)

S. K. Dutcher, L. Li, H. Lin, L. Meyer, T.H. Giddings, A.L. Kwan, B.L. Lewis, Whole-genome sequencing to identify mutants and polymorphisms in *Chlamydomonas reinhardtii*. G3.Genes|Geonmes|Genetics **2**,15–22 (2012)

J. Edzwald, Algae, bubbles, coagulants, and dissolved air flotation. Water Sci. Technol. **27**, 67–81 (1993)

J. Fan, Y. Cui, M. Wan, W. Wang, Y. Li, Lipid accumulation and biosynthesis genes response of the oleaginous *Chlorella pyrenoidosa* under three nutrition stressors. Biotechnol. Biofuels **7**, 17 (2014)

Z. Fengwu, B. Xinqing, Construction of Transgenic Flocculation Microalgae and Application Thereof in Microalgae Recovery, WO2013185388 (2013)

M. Fuhrmann, W. Oertel, P. Hegemann, A synthetic gene coding for the green fluorescent protein (GFP) is a versatile reporter in *Chlamydomonas reinhardtii*. Plant J. **19**, 353–361 (1999)

A. Ghosh, S. Khanra, M. Mondal, G. Halder, O.N. Tiwari, S. Saini, T.K. Bhowmick, K. Gayen, Progress toward isolation of strains and genetically engineered strains of microalgae for production of biofuel and other value added chemicals: a review. Energy Convers. Manag. 104–118 (2016)

J.A. Gimpel, V. Henríquez, S.P. Mayfield, In metabolic engineering of eukaryotic microalgae: potential and challenges come with great diversity. Front. Microbiol. **6**, 1376 (2015)

M.A. Gomaa, L. Al-Haj, R.M.M. Abed, Metabolic engineering of Cyanobacteria and microalgae for enhanced production of biofuels and high-value products. J. Appl. Microbiol. **121**(4), 919–931 (2016)

D. González-Ballester, A. deMontaigu, J.J. Higuera, A. Galvan, E. Fernandez, Functional genomics of the regulation of the nitrate assimilation pathway in Chlamydomonas. Plant Physiol. **137**, 522–533 (2005)

D. González-Ballester, W. Pootakham, F. Mus, W. Yang, C. Catalanotti, L. Magneschi, A. deMontaigu, J.J. Higuera, M. Prior, A. Galván, E. Fernandez, A.R. Grossman, Reverse genetics in Chlamydomonas: a platform for isolating insertional mutants. Plant Methods **7**, 24 (2011)

D.R. Georgianna, M.M. Hannon, S. Wu, K. Botsch, A.J. Lewis, J. Hyun, M. Mendez, S.P. Mayfield, Production of recombinant enzymes in the marine alga *Dunaliella tertiolecta*. Algal Res. **2**, 2–9 (2013)

H.C. Greenwell, L.M.L. Laurens, R.J. Shields, R.W. Lovitt, K.J. Flynn, Placing microalgae on the biofuels priority list: a review of the technological challenges. J. ı. Soc. Interface **7**, 703–726 (2010)

M. Hannon, J. Gimpel, M. Tran, B. Rasala, S. Mayfield, Biofuels from algae: challenges and potential. Biofuels **1**, 763–784 (2010)

Q. Hu, M. Sommerfeld, E. Jarvis, M. Ghirardi, M. Posewitz, M. Seibert, A. Darzins, Microalgal triacylglycerols as feedstocks for biofuel production: perspectives and advances. Plant J. **54**(4), 621–639 (2008)

II.C. Huppe, D.H. Turpin, Integration of carbon and nitrogen metabolism in plant and algal cells. Annu. Rev. Plant Biol. **45**, 577–607 (1994)

M. Isleten-Hosoglu, I. Gultepe, M. Elibol, Optimization of carbon and nitrogen sources for biomass and lipid production by *Chlorella saccharophila* under heterotrophic conditions and development of Nile red fluorescence based method for quantification of its neutral lipid content. Biochem. Eng. J. **61**, 11–19 (2012)

B.B. Jeong, D. Wu-Scharf, C. Zhang, H. Cerutti, Suppressors of transcriptional transgenic silencing in Chlamydomonas are sensitive to DNA-damaging agents and reactivate transposable elements. Proc. Natl. Acad. Sci. U S A **99**, 1076–1081 (2002)

B.K. Kaiser, M. Carleton, J.W. Hickman, C. Miller, D. Lawson, M. Budde, P. Warrener, A. Paredes, S. Mullapudi, P. Navarro, F. Cross, J.M. Roberts, Fatty aldehydes in cyanobacteria are a metabolically flexible precursor for a diversity of biofuel products. PLoS ONE **8**, 58307 (2013)

M. Kamalanathan, P. Chaisutyakorn, R. Gleadow, J. Beardall, A comparison of photoautotrophic, heterotrophic, and mixotrophic growth for biomass production by the green alga *Scenedesmus* sp. (Chlorophyceae). Phycologia **57**, 309–317 (2017)

F. Khademi, I. Yıldız, A.C. Yildiz, S. Abachi, Advances in algae harvesting and extracting technologies for biodiesel production, in: *Progress in Clean Energy*, vol. 2 (Springer International Publishing, Cham, 2015), pp. 65–82

M.I. Khan, J.H. Shin, J.D. Kim, The promising future of microalgae: current status, challenges, and optimization of a sustainable and renewable industry for biofuels, feed, and other products. Microb. Cell Fact. **17**, 36 (2018)

S. Kumar, GM algae for biofuel production: biosafety and risk assessment. Collect Biosaf. Rev. **9**, 52–75 (2015)

T. Kusakabe, T. Tatsuke, K. Tsuruno, Y. Hirokawa, S. Atsumi, J.C. Liao, T. Hanai, Engineering a synthetic pathway in cyanobacteria for isopropanol production directly from carbon dioxide and light. Metab. Eng. **20**, 101–108 (2013)

K.Y. Kwa, P.H. Jin, L.S. Hoon, L.J. Hwa, H.S. Keun, P.H. Young, Novel *Arthrospira platensis* Having High Flocculation Activity, KR20130094382 (2013)

M. Lapuerta, J. Barba, A.D. Sediako, M.R. Kholghy, M.J. Thomson, Morphological analysis of soot agglomerates from biodiesel surrogates in a coflow burner. J. Aerosol. Sci. **111**, 65–74 (2017)

Z. Li, L. Qiu, X. Cheng, Y. Li, H. Wu, The evolution of soot morphology and nanostructure in laminar diffusion flame of surrogate fuels for diesel. Fuel **211**, 517–528 (2018)

J. Lowrey, M.S. Brooks, P.J. McGinn, Heterotrophic and mixotrophic cultivation of microalgae for biodiesel production in agricultural wastewaters and associated challenges—a critical review. J. Appl. Phycol. **27**, 1485–1498 (2015)

X. Liu, J. Sheng, R. Curtiss, Fatty acid production in genetically modified cyanobacteria. Proc. Natl. Acad. Sci. U S A **108**, 6899–6904 (2011)

F. Martinez, C. Ascaso, M.I. Orus, Morphometric and stereologic analysis of *Chlorella vulgaris* under heterotrophic growth conditions. Ann. Bot. **67**, 239–245 (1991)

T.M. Mata, R. Almeidab, N.S. Caetanoa, Effect of the culture nutrients on the biomass and lipid productivities of microalgae *Dunaliella tertiolecta*. Chem. Eng. **32**, 973 (2013)

T.M. Mata, A.A. Martins, N.S. Caetano, Microalgae for biodiesel production and other applications: a review. Renew. Sustain. Energy Rev. **14**, 217–232 (2010)

T. Mathimani, N. Mallick, A comprehensive review on harvesting of microalgae for biodiesel—key challenges and future directions. Renew. Sustain. Energy Rev. **91**, 1103–1120 (2018)

S.R. Medipally, F.M. Yusoff, S. Banerjee, M. Shariff, Microalgae as sustainable renewable energy feedstock for biofuel production. BioMed. Res. Int. **2015** (Article Number: 519513) (2015)

K. Miyake, K. Abe, S. Ferri, M. Nakajima, M. Nakamura, W. Yoshida, K. Kojima, K. Ikebukuro, K. Sode, A green-light inducible lytic system for cyanobacterial cells. Biotechnol. Biofuels **7**(1), 1–8 (2014)

K. Muylaert, D. Vandamme, B. Meesschaert, I. Foubert, Flocculation of microalgae using cationic starch, in *Phycologia. The Physiological Society* (London, 2009), p. 63

J. Neupert, D. Karcher, R. Bock, Generation of Chlamydomonas strains that efficiently express nuclear transgenes. Plant J. **57**, 1140–1150 (2009)

M.E. Ngo-mati, C.A. Pieme, M. Azabjikenfack, B.M. Moukette, E. Korosky, P. Stefanini, J.Y. Ngogang, C.M. Mbofung, Impact of daily supplementation of *S. platensis* on the immune system of naive HIV-1 patients in Cameroon: a 12-months single blind, randomized, multicenter trial. Nutrition J. **14**, 2–7 (2015)

E. Nwokoagbara, A.K. Olaleye, M. Wang, Biodiesel from microalgae: the use of multi-criteria decision analysis for strain selection. Fuel**159**, 241–249 (2015)

J.W.K. Oliver, L.M.P. Machado, H. Yoneda, S. Atsumi, Combinatorial optimization of cyanobacterial 2,3-butanediol production. Metab. Eng. **22**, 76–82 (2014)

S.L. Pahl, A.K. Pahl, T. Kalaitzidis, P.J. Ashman, S. Sathe, D.M. Lewis, Harvesting, thickening and dewatering microalgae biomass, in *Algae for Biofuels and Energy*, ed. by M. Borowitzka, N. Moheimani (Springer, Berlin, 2013), pp. 165–185

M.B. Pereira, B.M.E. Chagas, R. Sassi, G.F. Medeiros, E.M. Aguiar, L.H.F. Borba, J.C. Andrade Neto, A.H.N. Rangel, Mixotrophic cultivation of *Spirulina platensis* in dairy wastewater: Effects on the production of biomass, biochemical composition and antioxidant capacity. PLoS ONE **14**(10), e0224294.42 (2019)

P. Přibyl, V. Cepak, V. Zachleder, Production of lipids in 10 strains of *Chlorella* and *Parachlorella*, and enhanced lipid productivity in *Chlorella vulgaris*. Appl. Microbiol. Biotechnol. **94**, 549–561 (2012)

B. Pushparaj, E. Pelosi, G. Torzillo, R. Materassi, Microbial biomass recovery using a synthetic cationic polymer. Bioresour. Technol. **43**, 59–62 (1993)

R. Radakovits, R.E. Jinkerson, A. Darzins, M.C. Posewitz, Genetic engineering of algae for enhanced biofuel production. Eukaryot. Cell **9**, 486–501 (2010)

B. Ravindran, M.B. Kurade, A.N. Kabra, B.H. Jeon, S.K. Gupta, Recent advances and future prospects of microalgal lipid biotechnology, in *Algal Biofuels* (Springer, 2017), pp. 1–37

I.M. Remmers, R.H. Wijffels, M.J. Barbosa, P.P. Lamers, Can we approach theoretical lipid yields in microalgae? Trends Biotechnol. **36**, 265–276 (2018)

A. Robles-Medina, P.A. Gonzalez-Moreno, L. Esteban-Cerdan, E. Molina-Grima, Biocatalysis: towards ever greener biodiesel production. Biotechnol. Adv. **27**, 398–408 (2009)

M.Y. Roleda, S.P. Slocombe, R.J.G. Leakey, J.G. Day, E.M. Bell, M.S. Stanley, Effects of temperature and nutrient regimes on biomass and lipid production by six oleaginous microalgae in batch culture employing a two-phase cultivation strategy. Biores. Technol. **129**, 439–449 (2013)

S. Roy, R. Pal, Microalgae in aquaculture: a review with special references to nutritional value and fish dietetics. Proc. Zool. Soc. **68**, 1–8 (2015)

A.M. Ruffing, Borrowing genes from *Chlamydomonas reinhardtii* for free fatty acid production in engineered cyanobacteria. J. Appl. Phycol. **25**(5), 1495–1507 (2013)

A.M. Ruffing, H.D.T. Jones, Physiological effects of free fatty acid production in genetically engineered *Synechococcus elongatus* PCC 7942. Biotechnol. Bioeng. **109**, 2190–2199 (2012)

B. Sajjadi, W.Y. Sajjadi, A.A.A. Raman, S. Raman, Microalgae lipid and biomass for biofuel production: a comprehensive review on lipid enhancement strategies and their effects on fatty acid composition. Renew. Sustain. Energy Rev. **97**, 200–232 (2018)

C.J. Salas-Montantes, O. González-Ortega, A.E. Ochoa-Alfaro, R. Camarena-Rangel, L.M.T. Paz-Maldonado, S. Rosales-Mendoza, A. Rocha-Uribe, R.E. Soria-Guerra, Lipid accumulation during nitrogen and sulfur starvation in *Chlamydomonas reinhardtii* overexpressing a transcription factor. J. Appl. Phycol. **30**(3), 1–13 (2018)

P. Savakis, K.J. Hellingwerf, Engineering cyanobacteria for direct biofuel production from CO_2. Curr. Opin. Biotechnol. **33**, 8–14 (2015)

M.G. Saad, N.S. Dosoky, M.S. Zoromba, H.M. Shafik, Algal biofuels: current status and key challenges. Energies **12**(10), 1920 (2019). https://doi.org/10.3390/en12101920

M. Scarsella, G. Belotti, P. De Filippis, M. Bravi, Study on the optimal growing conditions of *Chlorella vulgaris* in bubble column photobioreactors. Chem. Eng. **20**, 85–90 (2010)

P.M. Schenk, S.R. Thomas-Hall, E. Stephens, U.C. Marx, J.H. Mussgnug, C. Posten, O. Kruse, B. Hankamer, Second generation biofuels: high-efficiency microalgae for biodiesel production. Bioenergy Res. **1**, 20–43 (2008)

J.H. Shah, A. Deokar, K. Patel, K. Panchal, V. Alpesh, A.V. Mehta, A comprehensive overview on various method of harvesting microalgae according to Indian, in *Perspective, International Conference on Multidisciplinary Research & Practice*, vol. I, Issue VII (2017), p. 313. IJRSI. ISSN 2321-2705

R. Sharon-Gojman, S. Leu, A. Zarka, Antenna size reduction and altered division cycles in self-cloned, marker-free genetically modified strains of *Haematococcus pluvialis*. Algal Res. **28**, 172–183 (2017)

C.R. Shen, J.C. Liao, Photosynthetic production of 2-methyl-1-butanol from CO_2 in cyanobacterium *Synechococcus elongatus* PCC7942 and characterization of the native acetohydroxyacid synthase. Energy Environ. Sci. **5**, 9574–9583 (2012)

S.V. Shestakov, N.T. Khyen, Evidence for genetic transformation in blue-green alga *Anacystis nidulans*. Mol. Genet. Genomics **107**, 372–375 (1970)

Y.S. Shin, H.I. Choi, J.W. Choi, J.S. Lee, Y.J. Sung, S.J. Sim, Multilateral approach on enhancing economic viability of lipid production from microalgae: a review. Bioresour. Technol. **258**, 335–344 (2018)

Z. Shokravi, H. Shokravi, M.M.A. Aziz, H. Shokravi, The fourth-generation biofuel: a systematic review on nearly two decades of research from 2008 to 2019, in *Fossil Free Fuels Trends Renewable Energy* ed. by M.M.B.A. Aziz (Taylor and Francis, London, 2019)

P.L. Show, M.S.Y. Tang, D. Nagarajan, T.C. Ling, C.W. Ooi, J.S. Chang, A holistic approach to managing microalgae for biofuel applications. Int. J. Mol. Sci. **18**, 215 (2017)

M. Siaut, S. Cuine, C. Cagnon, B. Fessler, M. Nguyen, P. Carrier, A. Beyly, F. Beisson, C. Triantaphylides, Y. Li-Beisson, G. Peltier, Oil accumulation in the model green alga *Chlamydomonas reinhardtii*: characterization variability between common laboratory strains and relationship with starch reserves. BMC Biotechnol. **11**, 7 (2011). https://doi.org/10.1186/1472-6750-11-7

G. Singh, S.K. Patidar, Microalgae harvesting techniques: a review. J. Environ. Manage. **217**, 499–508 (2018)

N.K. Singh, A.K. Upadhyay, U.N. Rai, Algal technologies for wastewater treatment and biofuels production: an integrated approach for environmental management, in *Algal Biofuels* ed. by S. Gupta, A. Malik F. Bux (Springer, Berlin, 2017)

D.L. Sutherland, C. Howard-Williams, M.H. Turnbull, P.A. Broady, R.J. Craggs, Enhancing microalgal photosynthesis and productivity in wastewater treatment high rate algal ponds for biofuel production. Biores. Technol. **184**, 222–229 (2015)

E.B. Sydney, A.C. Novak Sydney, J.S. de Carvalho, C.R. Soccol, Microalgal strain selection for biofuel production, in *Biofuels from Algae* (Elsevier, 2019)

S.J. Szyjka, S. Mandal, N.G. Schoepp, B.M. Tyler, C.B. Yohn, Y.S. Poon, S. Villarealb, M.D. Burkarta, J.B. Shurinb, S.P. Mayfield, Evaluation of phenotype stability and ecological risk of a genetically engineered alga in open pond production. Algal Res **24**, 378–386 (2017)

A. Tabernero, E.M.M. del Valle, M.A. Galan, Evaluating the industrial potential of biodiesel from a microalgae heterotrophic culture: Scale-up and economics. Biochem. Eng. J. **63**, 104–115 (2012)

Y.J. Tamayo-Ordóñez, B.A. Ayil-Gutiérrez, F.L. Sánchez-Teyer, E.A. De la Cruz-Arguijo, F.A. Tamayo-Ordóñez, A.V. Córdova-Quiroz, M.C. Tamayo-Ordóñez, Advances in culture and genetic modification approaches to lipid biosynthesis for biofuel production and in silico analysis of enzymatic dominions in proteins related to lipid biosynthesis in algae. Phycol. Res. **65**, 14–28 (2017)

C.L. Teo, M. Atta, A. Bukhari, M. Taisir, A.M. Yusuf, A. Idris, Enhancing growth and lipid production of marine microalgae for biodiesel production via the use of different LED wavelengths. Biores. Technol. **162**, 38–44 (2014)

D. Vandamme, I. Foubert, K. Muylaert, Flocculation as a low-cost method for harvesting microalgae for bulk biomass production. Trends Biotechnol. **31**(4), 233–239 (2013)

A.M. Varman, Y. Xiao, H.B. Pakrasi, Y.J. Tang, Metabolic engineering of Synechocystis 6803 for isobutanol production. Appl. Environ. Microbiol. **79**, 908–914 (2013)

B. Wang, Y. Li, N. Wu, C. Lan, CO_2 bio-mitigation using microalgae. Appl. Microbiol. Biotechnol **79**, 707–718 (2008)

W. Wang, X. Liu, X. Lu, Engineering cyanobacteria to improve photosynthetic production of alka(e)nes. Biotechnol. Biofuels **6**, 69 (2013)

J. Wang, H. Yang, F. Wang, Mixotrophic cultivation of microalgae for biodiesel production: status and prospects. Appl. Biochem. Biotechnol.**172**, 3307–3329 (2014)

L. Xin, H. Hong-ying, G. Ke, S. Ying-xue, Effects of different nitrogen and phosphorus concentrations on the growth, nutrient uptake, and lipid accumulation of a freshwater microalga *Scenedesmus* sp. Biores. Technol. **101**, 5494–5500 (2010)

J. Xue, Y.F. Niu, T. Huang, W.D. Yang, J.S. Liu, H.Y. Li, Genetic improvement of the microalga *Phaeodactylum tricornutum* for boosting neutral lipid accumulation. Metab. Eng. **27**, 1–9 (2015)

C. Yeesang, B. Cheirsilp, Effect of nitrogen, salt, and iron content in the growth medium and light intensity on lipid production by microalgae isolated from freshwater sources in Thailand. Biores. Technol. **102**, 3034–3040 (2011)

J. Zhan, J. Rong, Q. Wang, Mixotrophic cultivation, a preferable microalgae cultivation mode for biomass/bioenergy production, and bioremediation, advances and prospect. Int. J. Hydrogen Energy **42**(12), 8505–8517 (2017)

B. Zhu, G. Chen, X. Cao, D. Wei, Molecular characterization of CO_2 sequestration and assimilation in microalgae and its biotechnological applications. Biores. Technol. **244**, 1207–1215 (2017)

N.N. Zulu, J. Popko, K. Zienkiewicz, P. Tarazona, C. Herrfurth, I. Herrfurth, Heterologous co-expression of a yeast diacylglycerol acyltransferase (ScDGA1) and a plant oleosin (AtOLEO3) as an efficient tool for enhancing triacylglycerol accumulation in the marine diatom *Phaeodactylum tricornutum*. Biotechnol. Biofuels, **10** (Article Number: 187) (2017)

Further Reading

A.A. Snow, Val H. Smith, Genetically engineered algae for biofuels: a key role for ecologists. BioScience **62**(8), 765–768 (2012)

S. Behera, R.C. Mohanty, R.C. Ray, Batch ethanol production from cassava (*Manihotesculenta crantz*) flourusing *Saccharomyces cerevisiae* cells immobilized in calcium alginate. Ann. Microbiol. **65**, 779–783 (2014)

I. Pereira, A. Rangel, B. Chagas, B. de Moura, S. Urbano, R. Sassi, F. Camara, C. Castro, Microalgae growth under mixotrophic condition using agro-industrial waste: a review. IntechOpen (2021). https://doi.org/10.5772/intechopen.93964

Chapter 3
Residue from Biofuel Extraction

The merits of microalgae as a source of biofuel feedstock have been broadly recognized as mentioned in Chap. 1. Benefits include production using non-arable land and brackish water which are not used in food production, potential recovery of waste nutrients from water treatment, and reduction of greenhouse gas emissions. Also, several non-biofuel-related uses of microalgae have been identified (Bryant et al. 2012). These are used as a livestock or aquaculture feed ingredient and used in the production of high-value oils for nutritional and pharmaceuticals supplements for people and animals, biotechnological applications, cosmetics, pigments, and agrochemicals (Chisti 2008; Brune et al. 2009; Lundquist et al. 2010; Stephens et al. 2010; Wijffels and Barbosa, 2010; Becker, 2007; Spolaore et al. 2006; Waltz, 2009; Sankaran et al. 2018). Up to 70% of the whole microalgae biomass can be obtained as residues after they serve their primary purpose.

The residue obtained after extracting the algal biomass is valuable as it is rich in nutritional components. The residues generated in biofuel production can be used for animal consumption in several industries for instance fish aquaculture, poultry, and cattle industries. Algal co-products can be used not only as a source of animal feed, but they also promote environmentally friendly technology (Van Asselt et al. 2011).

Patterson and Gatlin (2013) replaced fishmeal and soy protein concentration in the diet of *Sciaenops ocellatus* with lipid-extracted *Chlorella* sp. Spent microalgal biomass could replace 5% to 25% of crude protein in the reference diet. Replacement of 10% of dietary protein with spent microalgal biomass was found to be efficient as no reduction in fish growth was observed. Integration of LEA into aquaculture diets could possibly reduce the feed prices.

Drewery (2012) showed the potential of spent microalgal biomass to be used as animal feed. Incorporation of post-extraction algal residue as a protein source in beef cattle was found to be feasible. This will improve the economic sustainability of biofuel production from microalgae.

Algal biomass mainly contains carbohydrates, lipids, proteins, nucleic acid, and inorganic compounds. The solid material extracted from recycling culture media after the processing can be used as a secondary source of nutrients in cultivation systems (Moller and Muller, 2012). The use of the recycled biomass residue of anaerobic decomposition in biogas production has been studied. Digestate of anaerobic digestion is found to be rich in mineralized nutrients such as nitrogen and phosphorus (Alburquerque et al. 2012).

Thermal treatment is widely used for degrading DNA in genetically modified microorganisms.

Zheng et al. (2012) performed the enzymatic hydrolysis of the spent microalgal biomass of *Chlorella vulgaris* after lipid extraction and used it as a nutrient source for growing the same microalgae under aerated and nonaerated conditions. The better yield was obtained with the aerated cultures. The biomass production, lipid content, and lipid productivity were 3.28 g/l, 35%, and 116 mg/l/ day, respectively.

Gao et al. (2012) used *Pseudochoricystis ellipsoidea* which is an oleaginous microalgae for the production of microalgae. They analyzed the spent microalgal biomass after lipid extraction. It was found to be rich in proteins. Hydrolysis of proteins was conducted using a thermal-acid method. The hydrolysate was used to replace the yeast extract in lactic acid and ethanol fermentation. The addition of yeast extract was found to be essential for the efficient production of lactic acid. The ethanol fermenting yeast also needed yeast extract, but the dose requirement was lower.

Talukder et al. (2012) separated lipids by treating *Nannochloropsis salina* biomass by acid hydrolysis followed by solvent extraction. The lipid-free extract was used as a nutrient source in lactic acid fermentation. A lactic acid yield of 92.8% was achieved with the addition of 3 to 25 g/l of the microalgae-reducing sugars.

Anaerobic digestion of spent microalgal biomass yields methane and fermentation could produce biohydrogen (Lv et al. 2008). Their production is controlled by many elements which include upstream processes, pretreatment, protein content, nature of the primary products, and C/N ratio.

Yang et al. (2011) used spent microalgal biomass of *Scenedesmus* sp. after extraction of the oil for the production of biohydrogen. The heat-treated anaerobic sludge was used as inoculum. Several pretreatment methods were examined. The optimal hydrogen production with a hydrogen production rate of 2.82 mL/h and hydrogen yield of 30.03 m/g volatile solids was achieved.

Venkata Subhash and Venkata Mohan (2014) used mixed microalgal consortia for the extraction of lipids and the spent microalgal biomass was pretreated using the hydrothermal acid method for releasing the reducing sugars. Untreated spent microalgal biomass and treated spent microalgal biomass (extract and solids) and treated algal extract and solids were used for producing hydrogen with mixed acidogenic consortia. Maximum hydrogen production of 4.9 mol kg/ COD was obtained for the pretreated algal extract.

The spent microalgal biomass rich in sugars can be used as a feedstock for bioethanol fermentation. Saccharification is an important step for converting macromolecules of carbohydrates into fermentable sugars for ethanol production. Saccharification of spent microalgal biomass is much easier as compared to other feedstocks. *N. salina* lipid-extracted biomass was used to produce biomass (Mendez et al. 2014).

The deoiled biomass of *Scenedesmus* sp. CCNM 1077 was found to be rich in carbohydrates (45% DW). The spent microalgal biomass was pretreated by the chemoenzymatic method to produce reducing sugars with a 43% saccharification yield (Pancha et al. 2016). The spent microalgal biomass hydrolysate was used for ethanol fermentation with 78% ethanol production efficiency.

Chng et al. (2016) studied the sustainable production of bioethanol using lipid-extracted biomass from *Scenedesmus dimorphus*. The lipid-extracted biomass was directly subjected to simultaneous saccharification and fermentation, thus, avoiding the costly pretreatment, reducing the contamination risk, and also reducing the problem of high sugar content. The overall conversion of more than 90% of the theoretical yield was obtained. The maximum bioethanol yield was 0.26 g bioethanol/g lipid-extracted biomass.

Even though the algae are a relatively new and promising product in the market, the possible hazards of their consumption are mainly unknown. This situation may be due to the large variety of microalgal species and their different characteristics (Chisti, 2013). For better use of the remains of the biofuel production process as feed, the origin and nature of the microalgae should be specified. Concerns about feed safety risks are raised when there is not enough transparency regarding the origins of microalgae (Ng et al. 2017).

"Genetic engineering aims to enhance the production of algal biofuel. However, one of the major concerns of using GMOs is their disposal. The applied disposal methods should destruct the microorganism and the genetic element to minimize the risk of lateral gene transfer. DNA release in microorganisms generally occurs via cell lysis. The mechanisms of microorganisms remain active even after their death. This phenomenon may cause a release of plasmid or chromosomal DNA. The deliberate or accidental release of chromosomal or plasmid DNA at certain concentrations could result in horizontal gene transfer by transformation; hence, there are strict regulations for the disposal of these products. Waste reduction practices are aimed at minimizing the environmental effects associated with GM residue and at the replacement of hazardous by-products with more environmentally friendly alternatives. Several strategies could be applied to reduce waste (e.g., waste separation and concentration, waste reduction, energy/material recovery, waste exchange, incineration/treatment, and secure land disposal)" (Abdullah et al. 2019; Srivastava and Torres-Vargas 2017; Konig et al. 2004; Gogarten and Townsend 2005; Freeman 1988).

Composting offers a cost-effective method for disposing both genetically modified plant residues and genetically modified bacteria. "The conditions created in a properly managed composting process environment may help in destroying genetically modified organisms, and their genes, thereby reducing the risk of the spread of genetic material. When considering composting as a potential method for the disposal of genetically modified organisms, the establishment of controlled conditions providing an essentially homogenous environment appears to be important" (Singh et al. 2006).

The by-products of genetically engineered biomass must be decontaminated for preventing cross-pollination with wild species in surrounding ecosystems (Tsatsakis et al. 2017). Both the organism and the genetic material must be deconstructed concurrently to decontaminate genetically modified algae. The important factors in DNA degradation and in reducing the rate of horizontal transfer in genetically engineered organisms are temperature and pH. The DNA degradation is higher in compost than in soil because of the higher concentrations of microbes in active compost and subsequent increases in nucleases concentration (Torti et al. 2015).

Two classes of composting are mostly used for the degradation of genetically modified strains:

- Open turned windrow systems.
- Enclosed bioreactor systems.

Epstein (2017) recommends the use of an enclosed bioreactor because it significantly reduces the composting time and ensures that the materials are composted in an optimal condition. Reuse of the residues of algal biomass after processing its lipid, carbohydrate, and protein contents is a good and effective solution for removing concerns regarding the horizontal gene transfer (Heilmann et al. 2013).

Ueda et al. (1996) have suggested burning dried residue obtained from extracting the biofuel for recovering energy and carbon dioxide. A gas diffuser was used to add the carbon dioxide produced from the burning process to the culture medium. The energy obtained by the combustion of the dried residue can cover the operating energy of the process. Therefore, this method has three benefits: recovery of energy, recovery and reuse of carbon dioxide, and reduction of waste.

Huo et al. (2011) suggested extraction of the biofuel from the residue produced from the fermentation of algae. Genetically modified algae are one of the most favorable sources for the presented method as it allows the reuse of waste materials, involves the environment-friendly disposal of environmentally harmful remains, and achieves a maximum biofuel yield. Reuse of the algal residue in cultivation media is a solution that makes biofuel production sustainable by reducing the use of commercial fertilizers (Crofcheck and Crocker 2016).

While much research has documented the optimization of biomass utilization by improving processing conditions and media composition, not much research has studied the possibility of recycling growth medium after extraction of the energy (Lowrey et al. 2016).

Hrnčirova et al. (2008) studied the effects of thermal degradation on the quantity of the extracted DNA. Increasing the temperature substantially reduced the DNA content in genetically modified organisms in a time-dependent manner. The level of DNA degradation at a temperature of 200 °C within a period of four hours was substantially increased.

The effects of thermal processing on non-transgenic and transgenic DNA were studied by Bergerova et al. (2010). An important parameter in degradation was the size of the matrices of the extracted DNA. DNA in fine matrices was found to degrade faster. Thermal treatment might be an effective solution for DNA degradation; but the exposure time and temperature must be examined for each individual genetically modified algal strain (Beacham et al. 2017; Fernandes et al. 2016).

Abdullah et al. (2019) have compiled strategies used for waste management in the disposal of FGB production residues.

"Residue of the GM algae biomass obtained from energy extraction process may contain plasmid or chromosomal DNA that can be harmful to health and ecosystem. Hence, the treatment and handling of disposals must meet the requirements of the GM algae releasing regulations. Disposal methods that destruct the microorganism and the genetic elements are the most effective choices to minimize the risk of lateral gene transfer. The applied disposal method may have a direct impact on the economic viability and sustainability performance of the FGB. Saving energy can be achieved through burning the residue of the biofuel production process to cover the operating energy of the process. Based on the economic viability of FGB, the appropriate waste reduction method could be adapted. Water consumption during the production of FGB is another challenge that needs to be addressed to ensure the sustainable use of resources and biodiversity" (Abdullah et al. 2019).

References

B. Abdullah, S.A.F. Syed Muhammad, Z. Shokravi, S. Ismail, K.A. Kassim, A.N. Mahmood, M.M.A. Aziz, Fourth generation biofuel: a review on risks and mitigation strategies. Renew. Sustain. Energy Rev. **107**, 37–50 (2019)

J.A. Alburquerque, C. de la Fuente, A. Ferrer-Costa, L. Carrasco, J. Cegarra, M. Abad, M.P. Bernal, Assessment of the fertiliser potential of digestates from farm and agroindustrial residues. Biomass-Bioenergy **40**, 181–189 (2012)

T.A. Beacham, J.B. Sweet, M.J. Allen, Large scale cultivation of genetically modified microalgae: a new era for environmental risk assessment. Algal Res. **25**, 90–100 (2017)

E.W. Becker, Micro-algae as a source of protein. Biotechnol. Adv. **25**, 207–210 (2007)

E. Bergerova, Z. Hrnčirova, M. Stankovska, M. Lopašovska, P. Siekel, Effect of thermal treatment on the amplification and quantification of transgenic and non-transgenic soybean and maize DNA. Food Anal. Methods **3**, 211–218 (2010)

D.E. Brune, T.J. Lundquist, J.R. Benemann, Microalgal biomass for greenhouse gas reductions; potential for replacement of fossil-fuels and animal feeds. J. Environ. Eng. **135**, 1136–1144 (2009)

H.L. Bryant, I. Gogichaishvili, D. Anderson, J.W. Richardson, J. Sawyer, T. Wickersham, M.L. Drewery, The value of post-extracted algae residue. Algal Res. **1**(2), 185–193 (2012)

Y. Chisti, *Biodiesel from Microalgae Beats Bioethanol* (School of Engineering, Massey University, Private Bag 11 222, Palmerston North, New Zealand, 2008)

Y. Chisti, Constraints to commercialization of algal fuels. J. Biotechnol. **167**, 201–214 (2013)

L.M. Chng, D.J.C. Chan, K.T. Lee, Sustainable production of bioethanol using lipid-extracted biomass from *Scenedesmus dimorphus*. J. Clean. Prod. **130**, 68–73 (2016)

C. Crofcheck, M. Crocker, Application of recycled media and algae-based anaerobic digestate in Scenedesmus cultivation. J. Renew. Sustain. Energy **8**, 1 (2016)

M.L. Drewery, *Post-Extraction Algal Residue as a Protein Source For Cattle Consuming Forage* (Texas A&M University, Texas, USA, 2012)

E. Epstein, *The Science of Composting* (Routledge, New York, 2017)

T.J. Fernandes, J. Costa, A. Plácido, C. Villa, L. Grazina, L. Meira, M. Oliveira, I. Mafra, Genetically modified organism analysis as affected by DNA degradation. Genet. Modif. Org. Food Prod. Safety, Regul. Public Heal. 111–118 (2016)

H.M. Freeman, Hazardous waste minimization: a strategy for environmental improvement. JAPCA **38**, 59–62 (1988)

M.T. Gao, T. Shimamura, N. Ishida, H. Takahashi, Investigation of utilization of the algal biomass residue after oil extraction to lower the total production cost of biodiesel. J. Biosci. Bioeng. **114**(333), 330 (2012)

J.P. Gogarten, J.P. Townsend, Horizontal gene transfer, genome innovation and evolution. Nat. Rev. Microbiol. **3**, 679–687 (2005)

S.M. Heilmann, K.J. Valentas, M. Von Keitz, F.J. Schendel, P.A. Lefebvre, M.J. Sadowsky, L.A. Harned, L.R. Jader, Process for obtaining oils, lipids and lipid-derived materials from low cellulosic biomass materials, US PatentU2013/0206571 A1 **S20130206571A1**130206571A1 (2013)

Z. Hrnčirova, E. Bergerova, P. Siekel, Effects of technological treatment on DNA degradation in selected food matrices of plant origin. J. Food Nut. Res. **47** (2008)

Y.-X. Huo, K.M. Cho, J.G.L. Rivera, E. Monte, C.R. Shen, Y. Yan, J.C. Liau, Conversion of proteins into biofuels by engineering nitrogen flux. Nat. Biotechnol. **29**, 346 (2011)

A. Konig, A. Cockburn, R.W.R. Crevel, E. Debruyne, R. Grafstroem, U. Hammerling, I. Kimberg, I. Knudsen, H.A. Kuiper, A.A.C.M. Peijnenburgi, A.H. Penninks, M. Poulsen, M. Schauzuk, J.M. Wall, Assessment of the safety of foods derived from genetically modified (GM) crops. Food Chem. Toxicol. **42**, 1047–1088 (2004)

J. Lowrey, M.S. Brooks, R.E. Armenta, Nutrient recycling of lipid-extracted waste in the production of an oleaginous thraustochytrid. Appl. Microbiol. Biotechnol. **100**, 4711–4721 (2016)

T.J. Lundquist, I.C. Woertz, N.W.T. Quinn, J.R. Benjamin, A realistic technology and engineering assessment of algae biofuel production, in Technical report from the Energy Biosciences Institute (University of California, Berkeley, 2010)

P. Lv, C. Wu, L. Ma, Z. Yuan, A study on the economic efficiency of hydrogen production from biomass residues in China. Renew. Energy **33**, 1874–1879 (2008)

L. Mendez, A. Mahdy, M. Demuez, M. Ballesteros, C. González-Fernández, (2014). Effect of high pressure thermal pretreatment on *Chlorella vulgaris* biomass: organic matter solubilisation and biochemical methane potential. Fuel **117**, 674–679

K. Moller, T. Muller, Effects of anaerobic digestion on digestate nutrient availability and crop growth: a review. Eng. Life Sci. **12**, 242–257 (2012)

I.S. Ng, S.I. Tan, P.H. Kao, Y.K. Chang, J.S. Chang, Recent developments on genetic engineering of microalgae for biofuels and bio-based chemicals. Biotechnol J. **12** (2017)

I. Pancha, K. Chokshi, R. Maurya, S. Bhattacharya, P. Bachani, S. Mishra, Comparative evaluation of chemical and enzymatic saccharification of mixotrophically grown de-oiled microalgal biomass for reducing sugar production. Bioresour. Technol. **204**, 9–16 (2016)

D. Patterson, D.M. Gatlin, Evaluation of whole and lipid-extracted algae meals in the diets of juvenile red drum (*Sciaenops ocellatus*). Aquaculture **416–417**, 92–98 (2013)

R. Sankaran, P.L. Show, D. Nagarajan, J.S. Chang, Exploitation and biorefinery of microalgae, in *Waste Biorefinery* (Elsevier, 2018)

A. Singh, K. Billingsley, O. Ward, Composting: a potentially safe process for disposal of genetically modified organisms. Crit. Rev. Biotechnol. **26**, 1–16 (2006)

P. Spolaore, C. Joannis-Cassan, E. Duran, A. Isambert, Commercial applications of microalgae. J. Biosci. Bioeng. **101**(2), 87–96 (2006)

A. Srivastava, C.E. Torres-Vargas, Genetically modified crops: a long way to go. Environmental issues surroundings human overpopulation. IGI Glob. 104–119 (2017)

E. Stephens, I.I. Ross, Z. King, J.H. Mussgnug, O. Kruse, C. Posten, M.A. Borowitzka, B. Hankamer, An economic and technical evaluation of microalgal biofuels. Nat. Biotechnol. **28**(2), 126–128 (2010)

M.M.R. Talukder, P. Das, J.C. Wu, Microalgae (*Nannochloropsis salina*) biomass to lactic acid and lipid. Biochem. Eng. J. **68**, 109–113 (2012)

A. Torti, M.A. Lever, B.B. Jorgensen, Origin, dynamics, and implications of extracellular DNA pools in marine sediments. Mar. Genom. **24**, 185–196 (2015)

A.M. Tsatsakis, M.A. Nawaz, D. Kouretas, G. Balias, K. Savolainen, V.A. Tutelyan, K.S. Golokhvastbh, J. Dong, L.S. HwanYang, G. Chung, Environmental impacts of genetically modified plants: a review. Environ. Res. **156**, 818–833 (2017)

R. Ueda, S. Hirayama, K. Sugata, H. Nakayama, Process for the production of ethanol from microalgae, US Patent 5,578,472 (1996)

E.D. Van Asselt, M.P.M. Meuwissen, M. Van Asseldonk, P. Sterrenburg, M.J.B. Mengelers, H.J. Van der Fels-Klerx, Approach for a pro-active emerging risk system on biofuel by-products in feed. Food Policy **36**, 421–429 (2011)

G. Venkata Subhash, S. Venkata Mohan, Deoiled algal cake as feedstock for dark fermentative biohydrogen production: an integrated biorefinery approach. Int. J. Hydrog. Energy **39**, 9573–9579 (2014)

E. Waltz, Biotech's green gold? Nat. Biotechnol. **27**(1), 15–18 (2009)

R.H. Wijffels, M.J. Barbosa, An outlook on microalgal biofuels. Science **329**, 796–799 (2010)

Z. Yang, R. Guo, X. Xu, X. Fan, S. Luo, Fermentative hydrogen production from lipid-extracted microalgal biomass residues. Appl. Energy **88**, 3468–3472 (2011)

H. Zheng, Z. Gao, F. Yin, X. Ji, H. Huang, Lipid production of *Chlorella vulgaris* from lipid-extracted microalgal biomass residues through two-step enzymatic hydrolysis. Bioresour. Technol. **117**, 1–6 (2012)

Chapter 4
Environmental and Health Risks

Genetically engineered algae offer the promise of producing fuel, food, and other valuable products with reduced necessities for land and freshwater. While the gains in productivity measured in genetically engineered terrestrial crops are predicted to be mirrored in genetically engineered algae, the stability of phenotypes and ecological risks posed by genetically engineered algae in large-scale outdoor cultivation is not known (Szyjka et al. 2017).

For enhancing algal biomass, genetic engineering techniques have been used. These strategies are mostly based on the following (Rosenberg et al. 2008; Radakovits et al. 2010; Enzing et al. 2012):

- Target genes for the direct biosynthesis of biofuels.
- Lipid and carbohydrate metabolism.
- Feedstock secretion.
- Improved efficiency of nutrient use.
- Hydrogen production.
- Improved photosynthetic efficiency.
- Higher tolerance to stress.
- Enhanced cell disintegration.
- Bioflocculation.

These mechanisms can considerably improve the production of algal biofuels. But genetic modification cannot be relevant to all algal species mostly "due to a lack of available genomic data, the complexity of transgenesis, and difficulties establishing a competence equilibrium between metabolic and energy storage pathways. Enhancement of productivity and lipid accumulation is the easiest way to reduce the cost, nutrient consumption and water footprint. Genome editing methods are widely used for increasing the productivity and lipid content in microalgae. Presently, there are three types of genome editing tools: zinc-finger nuclease (ZFN), transcription activator-like effector nucleases (TALEN), and clustered regularly interspaced palindromic sequences (CRISPR/Cas9). The first genome editing experiment

P. Bajpai, *Fourth Generation Biofuels*,
SpringerBriefs in Applied Sciences and Technology,
https://doi.org/10.1007/978-981-19-2001-1_4

in microalgae was reported on *Chlamydomonas reinhardtii* using ZFN. Genome editing of *Phaeodactylum tricornutum* (with TALEN and with CRISPR/Cas9), *Nannochloropsis oceanica* (with CRISPR/Cas9), *C. reinhardtii* (with TALEN and CRISPR/Cas9) has been successfully demonstrated. The mechanisms of genome editing have been reviewed elsewhere" (Abdullah et al. 2019; Chu 2017; Maeda et al. 2018; Sizova et al. 2013; Serif et al. 2017; Stukenberg et al. 2018; Wang et al. 2016; Takahashi et al. 2018; Shin et al. 2019; Jeon et al. 2017).

High-throughput genetic engineering methods are becoming more and more efficient and economical and several microalgal strains have been examined for biofuel production. But genetically modified microalgae ecosystems because of their minute size, fast growth, and huge number can easily attack the ecosystems. The major environmental issues concerning the uncontained use of genetically modified algae are listed as follows (Hewett et al. 2016):

- Changes in natural habitats.
- Competition between the native species and the introduced microalgae and the native species.
- Horizontal gene transfer.
- Toxicity.

Non-native species invade more and multiply in native communities. So, the risk created by genetically modified microalgae mostly depends upon their endurance against abiotic and biotic stresses in the presence of native species (Szyjka et al. 2017). Many algal strains have been engineered to grow rapidly and stay alive in highly unfavorable environments. The risks must be addressed before these algal strains are dispersed from their open or controlled cultures (Adeniyi et al. 2018).

Release of the toxic algal strains into the environment creates severe risks to the health of humans (Assuncao et al. 2017). Issues regarding the environmental and health risks related to genetically modified algal strains were raised based on examples of intrusive aquatic species, like the multiplication of the toxic algae *Alexandrium minutum* in algae blooms along the French coast, which started in 1985 (Katsanevakis et al. 2014). Dinoflagellates can produce toxic compounds and biotoxins. The synthesis of toxic compounds is the main cause of algal blooms and the discoloration on sea surfaces. The damaging algal blooms events have been increasing all over the world over the last 40 years (Beacham et al. 2017). Horizontal gene transfer refers to one of several natural processes for the acquisition of genetic information via the stable transfer of genetic material from one distantly related organism to another outside of reproduction and without human intervention (Goldenfeld and Woese 2007). Cyanobacteria are considered important organisms for commercial applications because of their faster cell growth, higher potential for genetic modification, and simple requirement of nutrients. But the growth-monitoring of cyanobacteria is a difficult task because of the potential horizontal gene transfer between different cyanobacterial species and between cyanobacteria and eukaryotic algae. Moreover, some cyanobacterial species are able to take up DNA through viral vectors (Snow and Smith 2012).

The growing of genetically modified microalgae outdoor is critical to the sustainability and commercial production of FGB production (Sharma et al. 2011). But before introducing the genetically modified algae into the environment, risk assessments must be conducted for minimizing the potential safety and environmental issues associated with the flee of genetically modified traits from cultivation, harvesting, and processing facilities. Therefore, regulatory frameworks have been established in many countries to make certain that the use and processing of genetically modified algal biomass do not cause permanent harm to the environment through competition with the native species, changes in natural habitats, horizontal gene transfer, or toxicity.

According to Beachem et al. (2017) "Release of microalgae into the environment could have potential negative ecological effects such as altering food webs, displacing native phytoplankton, causing local extinctions, hazardous algal bloom formation, and having serious societal effects where harmful/toxic strains are involved. Many of the risks to human health and the environment associated with production of a given genetically modified microalgae will be specific to the types of traits and genes selected and the type of modifications performed. These genetically modified organisms specific risks should be considered alongside the risks of general large scale algae production and potential release into the environment. In addition to the specific traits associated with the genetically modified element of the microalgae other considerations will need to be made such as choice of algae (hazardous algal bloom formers or known invasive strains will have a higher associated risk), type and location of growth and containment facility, and the risk of horizontal gene transfer from the genetically modified algae to other organisms in the environment.

Many of the algae currently being modified are not native to the geographic areas in which they are generally cultivated and are often chosen for their rapid growth rate and overall hardiness which maximizes biomass productivity. Whilst there is currently very little regulatory control over the importation and release of non-native algal strains into the environment, such as in the use of microalgae in aquaculture, the risks associated with non-native invasion should also be considered. The actual environmental risk associated with large algae spills therefore will not be limited to the GM aspect of these organisms but rather a combination of factors including the fitness of the invading algae, the fitness of the indigenous alga populations, modes of competition for the resident and invading species, and intricacies and population stability characteristics of the disrupted ecological system. Indeed, since some transgenes reduce the fitness of recipient algae below the fitness of respective wild types, an important aspect of the risk analysis can therefore be based on the environmental risks associated with cultivating the wildtype" (Campbell 2011; Henley et al. 2013; Gressel et al. 2014).

Szyjka et al. (2017) described how genetically engineered algae perform in outdoor cultivation. This study was sanctioned by US EPA. *Acutodesmus dimorphus* was genetically engineered by adding two genes, one for improved fatty acid biosynthesis, and another for recombinant green fluorescence protein expression. Both the genes and their associated phenotypes were maintained during 50 days of outdoor cultivation. The genetically engineered algae dispersed from the cultivation ponds, but colonization of the trap ponds by the genetically engineered strain dropped fast with the increasing distance from the source cultivation ponds. On the contrary, several indigenous algae were found in every trap pond within a few days of starting the experiment. When inoculated in water from five local lakes, the effect of genetically engineered algae on biodiversity, species composition, and biomass of native algae was imperceptible from those of the wild-type progenitor algae, and neither the genetically engineered nor wild-type algae were able to outcompete native strains. The researchers concluded that genetically engineered algae can be grown successfully outdoors while maintaining genetically engineered traits, and that for the specific genetically engineered algal strain studied they did not outcompete or harmfully impact native algae when grown in water taken from local lakes. Uncontained growth of genetically engineered microalgae does not cause any unfavorable effect on the environment or the surrounding native algal population in the experimental period.

Henley et al. (2013) presented strategies for ecological risk assessment of genetically engineered algae for cultivation on commercial scale assuming that escape of genetically engineered algae into the environment cannot be avoided. They considered the potential ecological, economic, and health impacts of genetically engineered algae that persist in and change the natural ecosystems. Horizontal gene transfer with native organisms is of concern for certain traits, particularly when growing genetically engineered cyanobacteria. Most target genetically engineered algal traits are not likely to confer a selective benefit in nature, and therefore would rapidly reduce, resulting in lower ecological risk. Genetic and mechanical containment, plus conditional matching of genetically engineered algal traits to unnatural growth conditions, would reduce the risk further. These hypothetical predictions should be verified through thorough ongoing monitoring and mesocosm experiments for minimizing risk and fostering public and regulatory acceptance.

Table 4.1 summarizes the literature on the environmental and health risks associated with FGB. Table 4.2 shows the risks associated with genetically modified algae.

Table 4.1 Summary of the literature on the environmental and health risks of FGB

References	Cultivation	Harvesting	Processing
Van Asselt et al. (2011)	✓	X	X
Snow and Smith (2012)	✓	X	X
Menetrez (2012)	✓	X	X
Wijffels et al. (2013)	✓	X	X
Henley et al. (2013)	✓	X	X
Usher et al. (2014)	✓	X	X
Hegde et al. (2015)	✓	X	X

(continued)

Table 4.1 (continued)

References	Cultivation	Harvesting	Processing
Misra et al. (2016)	✓	X	X
De Farias Silva and Bertucco (2016)	✓	X	X
Zeraatkar et al. (2016)	✓	X	X
Mazard et al. (2016)	✓	X	X
Ghosh et al. (2016)	✓	X	X
Beacham et al. (2017)	✓	✓	✓
Chu (2017)	✓	x	x
Szyjka et al. (2017)	✓	X	X
Hess et al. (2018)	✓	X	X

Reproduced with permission from Abdullah et al. (2019)

Table 4.2 The health- and environment-related risk of GM algae

Topic	Risk contribution	References	Effect
Allergies	Human health	Menetrez (2012), Genitsaris et al. (2011), Mandel (2003)	Dermal, ingestive, respiratory exposure
Antibiotic resistance	Human health	Mandel (2003), Wright et al. (2013)	Reducing the effectiveness of medical treatments
Carcinogens	Human health	Menetrez (2012	Carcinogenic residues
Pathogenicity or toxicity	Human health	Snow and Smith (2012), Henley et al. (2013), Menetrez (2012)	Pathogenicity of some strain to human; toxic blooms; chemical transfer, toxic residues
Change or depletion of the environment	Environment	Tucker and Zilinskas (2006), Bhutkar (2005)	Removal of nutrients from ecosystem; reducing biodiversity of the flora and fauna
Competition with native species	Environment	Kuiken et al. (2014), Raybould (2010)	Outcompete native organisms; changing aquatic ecosystems
Horizontal gene transfer	Environment	Lu et al. (2010)	Transfer of genetic organisms
Pathogenicity or toxicity	Environment	Jeschke et al. (2013)	Pathogenicity of some strain to human; algal blooms; generating genetic-related toxins

Abdullah et al. (2019). Reproduced with permission

References

B. Abdullah, S.A.F. Syed Muhammad, Z. Shokravi, S. Ismail, K.A. Kassim, A.N. Mahmood, M.M.A. Aziz, Fourth generation biofuel: a review on risks and mitigation strategies. Renew. Sustain. Energy Rev. **107**, 37–50 (2019)

O.M. Adeniyi, U. Azimov, A. Burluka, Algae biofuel: current status and future applications. Renew. Sustain. Energy Rev. **90**, 316–335 (2018)

J. Assuncao, A. Guedes, F.X. Malcata, Biotechnological and pharmacological applications of biotoxins and other bioactive molecules from dinoflagellates. Mar. Drugs **15**, 393 (2017)

T.A. Beacham, J.B. Sweet, M.J. Allen, Large scale cultivation of genetically modified microalgae: a new era for environmental risk assessment. Algal Res. **25**, 90–100 (2017)

A. Bhutkar, Synthetic biology: navigating the challenges ahead. J Biolaw Bus **2005**(8), 19–29 (2005)

M.L. Campbell, Assessing biosecurity risk associated with the importation of nonindigenous microalgae. Environ. Res. **111**, 989–998 (2011)

W.L. Chu, Strategies to enhance production of microalgal biomass and lipids for biofuel feedstock. Eur J Phycol **52**, 419–437 (2017)

C.E. de Farias Silva, A. Bertucco, Bioethanol from microalgae and cyanobacteria: a review and technological outlook. Process Biochem **51**, 1833–1842 (2016)

C. Enzing, A. Nooijen, G. Eggink, J. Springer, R.H. Wijffels, Algae and genetic modification: research, production and risks. Research report, Technopolis Group, Amsterdam

S. Genitsaris, K.A. Kormas, M. Moustaka-Gouni, Airborne algae and cyanobacteria: occurrence and related health effects. Front. Biosci. **3**, 772–787 (2011)

A. Ghosh, S. Khanra, M. Mondal, G. Halder, O.N. Tiwari, S. Saini, Progress toward isolation of strains and genetically engineered strains of microalgae for production of biofuel and other value added chemicals: a review. Energy Convers. Manag. **113**, 104–118 (2016)

N. Goldenfeld, C. Woese, Biology's next revolution. Nature **445**, 369 (2007)

J. Gressel, C. Neal Stewart, L.V. Giddings, A.J. Fischer, J.C. Streibig, N.R. Burgos, A. Trewavas, A. Merotto, C.J. Leaver, K. Ammann, V. Moses, A. Lawton-Rauh, Overexpression of epsps transgene in weedy rice: insufficient evidence to support speculations about biosafety. New Phytol. **202**, 360–362 (2014)

K. Hegde, N. Chandra, S.J. Sarma, S.K. Brar, V.D. Veeranki, Genetic engineering strategies for enhanced biodiesel production. Mol. Biotechnol. **57**, 606–624 (2015)

W.J. Henley, R.W. Litaker, L. Novoveska, C.S. Duke, H.D. Quemada, R.T. Sayre, Initial risk assessment of genetically modified (GM) microalgae for commodity-scale biofuel cultivation. Algal Res. **2**, 66–77 (2013)

S.K. Hess, B. Lepetit, P.G. Kroth, S. Mecking, Production of chemicals from microalgae lipids—status and perspectives. Eur. J. Lipid Sci. Technol. **120** (2018)

J.P. Hewett, A.K. Wolfe, R.A. Bergmann, S.C. Stelling, K.L. Davis, Human health and environmental risks posed by synthetic biology R&D for energy applications: a literature analysis. Appl. Biosaf. **21**, 177–184 (2016)

S, Jeon, J.M. Lim, H.G. Lee, S.E. Shin, N.K. Kang, Y.I. Kang, Current status and perspectives of genome editing technology for microalgae. Biotechnol. Biofuels **10** (2017)

J.M. Jeschke, F. Keesing, R.S. Ostfeld, Novel organisms: comparing invasive species, GMOs, and emerging pathogens. Ambio **42**, 541–548 (2013)

S. Katsanevakis, I. Wallentinus, A. Zenetos, E. Leppakoski, M.E. Cinar, B. Ozturk, Impacts of invasive alien marine species on ecosystem services and biodiversity: a pan-European review. Aquat. Invasions **9**, 391–423 (2014)

T. Kuiken, G. Dana, K. Oye, D. Rejeski, Shaping ecological risk research for synthetic biology. J. Environ. Stud. Sci. **4**, 191–199 (2014)

S. Lu, L. Li, G. Zhou, Genetic modification of wood quality for second-generation biofuel production. GM Crops **1**, 230–236 (2010)

Y. Maeda, T. Yoshino, T. Matsunaga, M. Matsumoto, T. Tanaka, Marine microalgae for production of biofuels and chemicals. Curr. Opin. Biotechnol. **50**, 111–120 (2018)

G.N. Mandel, Gaps, inexperience, inconsistencies, and overlaps: crisis in the regulation of genetically modified plants and animals. William Mary Law Rev. **45**, 2167 (2003)

S. Mazard, A. Penesyan, M. Ostrowski, I.T. Paulsen, S. Egan, Tiny microbes with a big impact: the role of cyanobacteria and their metabolites in shaping our future. Mar. Drugs **14** (2016)

M.Y. Menetrez, An overview of algae biofuel production and potential environmental impact. Environ. Sci. Technol. **46**, 7073–7085 (2012)

N. Misra, P.K. Panda, B.K. Parida, B.K. Mishra, Way forward to achieve sustainable and cost-effective biofuel production from microalgae: a review. Int. J. Environ. Sci. Technol. **13**, 2735–2756 (2016)

R. Radakovits, R.E. Jinkerson, A. Darzins, M.C. Posewitz, Genetic engineering of algae for enhanced biofuel production. Eukaryot. Cell **9**, 486–501 (2010)

A. Raybould, The bucket and the searchlight: formulating and testing risk hypotheses about the weediness and invasiveness potential of transgenic crops. Environ. Biosaf. Res. **9**, 123–133 (2010)

J.N. Rosenberg, G.A. Oyler, L. Wilkinson, M.J. Betenbaugh, A green light for engineered algae: redirecting metabolism to fuel a biotechnology revolution. Curr. Opin. Biotechnol. **19**, 430–436 (2008)

M. Serif, B. Lepetit, K. Weisert, P.G. Kroth, C. Rio Bartulos, A fast and reliable strategy to generate TALEN-mediated gene knockouts in the diatom *Phaeodactylum tricornutum*. Algal Res **23**, 186–195 (2017)

N.K. Sharma, S.P. Tiwari, K. Tripathi, A.K. Rai, Sustainability and cyanobacteria (bluegreen algae): facts and challenges. J. Appl. Phycol. **23**, 1059–1081 (2011)

Y.S. Shin, J. Jeong, T.H.T. Nguyen, J.Y.H. Kim, E. Jin, S.J. Sim, Targeted knockout of phospholipase A$_2$ to increase lipid productivity in *Chlamydomonas reinhardtii* for biodiesel production. Bioresour. Technol. **271**, 368–374 (2019)

I. Sizova, A. Greiner, M. Awasthi, S. Kateriya, P. Hegemann, Nuclear gene targeting in Chlamydomonas using engineered zinc-finger nucleases. Plant J. **73**, 873–882 (2013)

A.A. Snow, V.H. Smith, Genetically engineered algae for biofuels: a key role for ecologists. Bioscience **62**, 765–768 (2012)

D. Stukenberg, S. Zauner, G. Dell'Aquila, U.G. Maier,. Optimizing CRISPR/cas9 for the diatom *Phaeodactylum tricornutum*. Front Plant. Sci. **9** (2018).

S.J. Szyjka, S. Mandal, N.G. Schoepp, B.M. Tyler, C.B. Yohn, Y.S. Poon, S. Villareal, M.D. Burkart, J.B. Shurin, S.P. Mayfield, Evaluation of phenotype stability and ecological risk of a genetically engineered alga in open pond production. Algal Res. **24**, 378–386 (2017)

K. Takahashi, Y. Ide, J. Hayakawa, Y. Yoshimitsu, I. Fukuhara, J. Abef et al., Lipid productivity in TALEN-induced starchless mutants of the unicellular green alga *Coccomyxa* sp. strain Obi. Algal Res. **32**, 300–307 (2018).

J.B. Tucker, R.A. Zilinskas, The promise and perils of synthetic biology. New Atlantis 25–45 (2006)

P.K. Usher, A.B. Ross, M.A. Camargo-Valero, A.S. Tomlin, W.F. Gale, An overview of the potential environmental impacts of large-scale microalgae cultivation. Biofuels **5**, 331–349 (2014)

E.D. Van Asselt, M.P.M. Meuwissen, M. Van Asseldonk, P. Sterrenburg, M.J.B. Mengelers, H.J. Van der Fels-Klerx, Approach for a pro-active emerging risk system on biofuel by-products in feed. Food Policy **2011**(36), 421–429 (2011)

Q. Wang, Y. Lu, Y. Xin, L. Wei, S. Huang, J. Xu, Genome editing of model oleaginous microalgae *Nannochloropsis* spp. by CRISPR/Cas9. Plant J. **88**, 1071–1081 (2016)

R.H. Wijffels, O. Kruse, K.J. Hellingwerf, Potential of industrial biotechnology with cyanobacteria and eukaryotic microalgae. Curr. Opin. Biotechnol. **24**, 405–413 (2013)

O. Wright, G.-B. Stan, T. Ellis, Building-in biosafety for synthetic biology. Microbiology **159**, 1221–1235 (2013)

A.K. Zeraatkar, H. Ahmadzadeh, A.F. Talebi, N.R. Moheimani, M.P. McHenry, Potential use of algae for heavy metal bioremediation, a critical review. J. Environ. Manag. **181**, 817–831 (2016)

Chapter 5
Regulations on Cultivation and Processing of Genetically Modified Algae

In general, it is accepted that the purposeful release of genetically modified organisms into the environment in most of the cases is an essential step in developing new products obtained from or containing genetically modified algae. These organisms whether released into the environment in smaller or large amounts may continue to exist, reproduce, and spread. The effects of such releases on the environment may be permanent (European Parliament and the Council of The European Union, Directive 2004/35/CE). Hence, before starting the production of genetically modified algae, an application should be submitted to the appropriate authorities for regulatory approval for releasing or marketing the algae and/or its derived products. These applications focus on a risk assessment which covers human health, environmental protection, labeling, and product use (Snow and Smith 2012). Furthermore, as public concerns might be the most important hurdle to commercialization of genetically modified algae (depending upon the type of product), handling of information and release must be engaging and transparent, and can be considered as part of, or in addition to, the risk assessment, to lessen the possibility of commercial failure because of rejection of product by consumers in response to the issues raised by activists.

Figure 5.1 shows the risk analysis decision support system: Factors to consider in relation to the "parent" wild type, the genetically modified algae, and the production life cycle (Beacham et al. 2017).

In recent years, there has been an increasing interest to develop genetically modified algae and other microorganisms for use in the production of biofuel and bio-based chemicals. But this comes at a time when there is a doubt within the scientific community as well as industry about how such uses will be regulated by governments in the United States and in other countries, and also the issues/concerns raised by some observers over the sufficiency of existing regulations for covering organisms produced using methods known as synthetic biology. But a reasonable road map is up-and-coming of a regulatory regime which allows pilot, demonstration, and commercial stage utilization of modified microorganisms.

© The Author(s), under exclusive license to Springer Nature Singapore Pte Ltd. 2022   49
P. Bajpai, *Fourth Generation Biofuels*,
SpringerBriefs in Applied Sciences and Technology,
https://doi.org/10.1007/978-981-19-2001-1_5

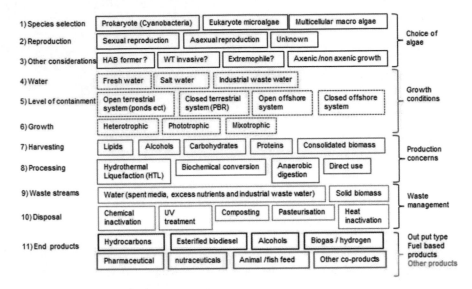

Fig. 5.1 Risk analysis decision support system: factors to consider in relation to the "parent" wild type, the GM algae, and the production life cycle (Beacham et al. 2017)

In the United States, regulations of the United States Environmental Protection Agency and possibly of the United States Department of Agriculture might govern the industrial use of microorganisms in closed photobioreactors or algae in open ponds, and these regulations normally need conducting assessments of the potential environmental risks of such large-scale use. The Environmental Protection Agency regulations include a mechanism by which outdoor utilization of genetically modified microorganisms can take place in a stepwise manner, with risks assessed as the scale of experimentation increases, which provides an accessible path to explore the use of genetically modified algae in open ponds. These types of risk assessments will address legitimate questions of potential ecological impact, such as the potential survival and dissemination of the production organism, the potential for heterologous genes to horizontally transfer to indigenous microbes, and the chance for other unintended effects on nontarget species. Several companies have effectively navigated these regulations, including some current project approvals in the United States and in other countries (Glass 2015).

The release of genetically modified algae into the environment is regulated due to the possible harm it causes to land water, protected species, and natural habitats (EC Directive 2004/35/EC). While the deliberate release of genetically modified algae into the environment is an important step for introducing new products, any release of genetically modified algae should be approved by the authorities before starting the production of genetically modified algae or their derivatives (EC. Directive 2004/35/CE). In many countries, clear regulations are in place for practices involving modified microbes which can be applied to algae without any uncertainty. Based on a ruling of the Environmental Protection Agency of the United States, any modified algal biomass is subject to the regulations under the Toxic Substances Control Act. The authorities should be informed in advance of any commercial, importation, or outdoor research and development (R&D) of the modified microorganism. Any non-commercial R&D activities performed under suitably contained conditions are not juried by the Toxic Substances Control Act. The potential impact of the large-scale uncontained cultivation of modified algae on agriculture or the environment is covered by "7 CFR Part 340" of the regulations outlined by the United States Department of Agriculture (Trentacoste et al. 2015). "The use of the modified algal organisms in a contained bioreactor is not within the scope of United States Department of Agriculture's regulations due to the low possibility that such practices will create any environmental hazards (Henley et al. 2013). The regulations on aquaculture, which are governed by the Food and Agriculture Organization, are applied in many countries. Wherever applicable, these regulations may lead to extra requirements in the way of permits or further environmental assessments (FAO 2009). The sanction on genetic modification of the European Union (EU) mainly relies on two directives (i.e., 2009/41/EC and 2001/18/EC) that pertain to the contained use of genetically modified microorganisms and the intended release of genetically modified organisms into the environment, respectively (Stellberger et al. 2017). The objective of Directive 2009/41/EC is to conserve the environment and protect human health through regulations on the contained use of genetically manipulated algae strains. Its primary focuses are to assess and classify the risks associated with genetic engineering and to provide containment (Fontana 2010; Smeets et al. 2014). In contrast, the regulation for the intended release of genetically modified organisms into the environment is stipulated in Directive 2001/18/EC and applies a step-by-step risk assessment approach (Garcia 2006).

The biotechnology safety directive regulated by the European Union is strict on the production, labelling, importing, and authorization of genetically modified algae products. The shipment of ingredients made from genetically modified algae or their by-products from the United States to Europe is presently not allowed. Therefore, these regulations have almost eliminated the European Union as a genetically modified organisms market" (Abdullah et al. 2019).

The EC directives on GMOs make a clear distinction between contained use and deliberate release into the environment (OECD 2015):

- Contained use is defined as "any activity in which organisms are genetically modified or in which such organisms are cultured, stored, transported, destroyed, disposed of or used in any other way and for which specific containment and other protective measures are used to limit their contact with the general public and the environment".
- Deliberate release is defined as "any intentional introduction into the environment of a GMO or a combination of GMOs for which no specific containment measures are used to limit their contact with, and to provide a high level of safety for, the general population and the environment".

Table 5.1 shows the regulatory sanctions for the production of FGB (Abdullah et al. 2019). The contained and uncontained use of genetically modified algae is subject

Table 5.1 Regulatory sanctioned for production of FGB

Glass (2015), Bergeson et al. (2014)	EPA regulations under TSCA	Exploitation
Trentacoste et al. (2015), Henley et al. (2013)	USDA biotechnology regulations	Exploitation and applicability
Glass (2015)	Food and Drug Administration (FDA) regulations	Applicability
EU-Commission (2001), Yamanouchi (2005), Darch and Shahsavarani (2012), Tribe (2012)	International biotechnology regulation	–
Glass (2015)	EPA regulation	Research uses
USDA (1993, 1995)	USDA regulations	Research uses
Glass (2015)	International regulation	Research uses
USEPA (1997)	EPA regulation	Commercial uses
Strauss et al. (2010)	USDA regulation	Commercial uses
FDA (2014)	FDA regulation	Animal feed uses
Takahashi et al. (2018)	International regulation	Commercial uses

Reproduced with permission from Abdullah et al. (2019)

to notifying and, in some cases, obtaining approval from the federal government of the country.

Under Japan's biotechnology regulatory regime, two types of uses for modified organisms ("Type I" and "Type II") are introduced. The "Type I" use is applicable to the deliberate release of modified organisms, which is different from the contained "Type II" use (Yamanouchi 2005).

The Australian regulation for the contained and uncontained use of modified microorganisms was implemented in 2011 under the name "Gene Technology Regulation" (Tribe 2012). A distinction is made between contained and uncontained uses of microorganisms, referred to as "dealings not involving release (DNIR)" and "dealings involving release (DIR)," respectively.

"Based on the Canadian Environmental Protection Act, the use of contained and uncontained modified microorganisms needs consent from the federal government. In response to the concerns raised by the potential environmental and health impacts of releasing modified biomass, government regulatory requirements are defined for contained and uncontained uses. The sanctions on the production and processing of these organisms are particularly stringent for both deliberate and accidental releases" (Abdullah et al. 2019; Darch and Shahsavarani 2012; Scaife et al. 2015).

The microalgae are generally grown in open ponds worldwide (Tan et al. 2018; Bharathiraja et al. 2015). But the use of open-pond reactors for growing genetically modified algae would pose additional hazards as compared with unmodified algal species. The field testing of these organisms has been suggested for assessing the potential risk of the growth and processing of the genetically modified microalgae on health and the environment. Within the coordinated framework of the EPA and USDA, the implementation of field tests and commercial uncontained production's environmental effects have been suggested (Glass 2015). Additionally, the EPA has sponsored projects for the open-environment utilization of algal biomass for assessing the attributed risks (Szyjka et al. 2017).

Environmental and health risks and also the public acceptance and attitude toward genetically modified products are the major driving forces for policy-makers for setting national targets and benchmarks for future development. Generally, the regulations on the discharge of genetically modified algae are divided into two most important classes, deliberate and unintended release. Industrial use of genetically modified algal biomass for FGB production requires clear regulations without any uncertainty for dealing with the deliberate or unintended release (Abdullah et al. 2019).

The sanctions on genetic modification in some countries such as the European Union are severe, and import of the genetically modified algae or their by-products is not permitted. Certain countries like the United States have less severe criteria to accept or reject a plan of production (Abdullah et al. 2019).

References

B. Abdullah, S.A.F. Syed Muhammad, Z. Shokravi, S. Ismail, K.A. Kassim, A.N. Mahmood, M.M.A. Aziz, Fourth generation biofuel: A review on risks and mitigation strategies. Renew. Sustain. Energy Rev. **107**, 37–50 (2019)

T.A. Beacham, J.B. Sweet, M.J. Allen, Large scale cultivation of genetically modified microalgae: a new era for environmental risk assessment. Algal Res. **25**, 90–100 (2017)

L.L. Bergeson, C.M. Auer, O. Hernandez, Creative adaptation: enhancing oversight of synthetic biology under the Toxic Substances Control Act. Ind. Biotechnol. **10**, 313–322 (2014)

B. Bharathiraja, M. Chakravarthy, R. Ranjith Kumar, D. Yogendran, D. Yuvaraj, J. Jayamuthunagai et al. Aquatic biomass (algae) as a future feed stock for biorefineries: a review on cultivation, processing and products. Renew. Sustain. Energy. Rev. **47**:634–53

H. Darch, A. Shahsavarani, The regulation of organisms used in agriculture under the Canadian Environmental Protection Act, 1999, in *Regulation of Agricultural Biotechnology* (Springer, The United States and Canada, 2012, pp. 137–45)

EC Directive 2004/35/EC of the European Parliament and of the Council of 21 April 2004 on environmental liability with regard to the prevention and remedying of environmental damage (ELD). Environ Liabil 2004

EU-Commission (2001). Directive 2001/18/EC of the European Parliament and of the Council of 12 March 2001 on the deliberate release into the environment of genetically modified organisms and repealing Council Directive 90/220. EEC 2001.

European Parliament and the Council of The European Union, Directive 2004/35/CE of the European parliament and of the Council of 21 April 2004 on environmentla liability with regard to the prevention and remedying of environmental damage. Off. J. Eur. Communities 1–23 (2004)

FAO, National aquaculture legislative overview-fact sheets (2009)

G. Fontana, Genetically modified micro-organisms: the EU regulatory framework and the new Directive 2009/41/EC on the contained use. Chem. Eng. Trans. **20**, 1–6 (2010)

Food US, Administration D, Generally recognized as safe (GRAS) notification program (2014)

P.R. Garcia, Directive 2001/18/EC on the deliberate release into the environment of GMOs: an overview and the main provisions for placing on the market. J. Eur. Environ. Plan. Law **3**, 3–12 (2006)

D.J. Glass, Pathways to obtain regulatory approvals for the use of genetically modified algae in biofuel or biobased chemical production. Ind. Biotechnol. **2015**(11), 71–83 (2015)

W.J. Henley, R.W. Litaker, L. Novoveska, C.S. Duke, H.D. Quemada, R.T. Sayre, Initial risk assessment of genetically modified (GM) microalgae for commodity-scale biofuel cultivation. Algal Res **2**, 66–77 (2013)

OECD, The need and risks of using transgenic micro-algae for the production of food, feed, chemicals and fuels, in *Biosafety and the Environmental Uses of Micro-Organisms: Conference Proceedings* (OECD Publishing, Paris, 2015). https://doi.org/10.1787/9789264213562-8-en

M.A. Scaife, A. Merkx-Jacques, D.L. Woodhall, R.E. Armenta, Algal biofuels in Canada: status and potential. Renew. Sustain. Energy Rev. **44**, 620–642.

E. Smeets, A. Tabeau, S. Van Berkum, J. Moorad, H. Van Meijl, G. Woltjer, The impact of the rebound effect of the use of first generation biofuels in the EU on greenhouse gas emissions: a critical review. Renew. Sustain. Energy. Rev. **2014**(38), 393–403 (2014)

A.A. Snow, V.H. Smith, Genetically engineered algae for biofuels: a key role for ecologists. Bioscience **62**, 765–768 (2012)

T. Stellberger, N. Koehler, A. Dinkelmeier, J. Draxler, M. Haase, J. Hellinckx, Strategies and methods for the detection and identification of viral vectors. Virus Genes **53**, 749–757 (2017)

S.H. Strauss, D.L. Kershen, J.H. Bouton, T.P. Redick, H. Tan, R.A. Sedjo, Far-reaching deleterious impacts of regulations on research and environmental studies of recombinant DNA-modified perennial biofuel crops in the United States. Bioscience **60**, 729–741 (2010)

S.J. Szyjka, S. Mandal, N.G. Schoepp, B.M. Tyler, C.B. Yohn, Y.S. Poon, S. Villareal, M.D. Burkart, J.B. Shurin, S.P. Mayfield, Evaluation of phenotype stability and ecological risk of a genetically engineered alga in open pond production. Algal Res. **24**, 378–386 (2017)

K. Takahashi, Y. Ide, J. Hayakawa, Y. Yoshimitsu, I. Fukuhara, J. Abe et al. (2018). Lipid productivity in TALEN-induced starchless mutants of the unicellular green alga *Coccomyxa* sp. strain Obi. Algal Res 2018;32:300–7.

X.B. Tan, M.K. Lam, Y. Uemura, J.W. Lim, C.Y. Wong, K.T. Lee, Cultivation of microalgae for biodiesel production: a review on upstream and downstream processing. Chin. J. Chem. Eng. **26**, 17–30 (2018)

E.M. Trentacoste, A.M. Martinez, T. Zenk, The place of algae in agriculture: policies for algal biomass production. Photosynth. Res. **123**, 305–315 (2015)

D. Tribe, Gene technology regulation in Australia: a decade of a federal implementation of a statutory legal code in a context of constituent states taking divergent positions. GM Crops. Food **3**, 21–29 (2012)

USDA, Plant health inspection service."genetically engineered organisms and products; simplification of requirements and procedures for genetically engineered organisms." Fed. Regist. **60**, 43567 (1995)

USDA Products, Notification procedures for the Introduction of certain regulated articles and Petition for Nonregulated status. 58. Fed. Regist. **17**, 44 (1993)

USEPA, *Points to Consider in The Preparation of TSCA Biotechnology Submissions for Microorganisms* (1997)

K. Yamanouchi, Regulatory considerations in the development and application of biotechnology in Japan. Rev. Sci. Tech. Int. Des. Epizooties **24**, 109 (2005)

Chapter 6
Carbon Dioxide Sequestration

The never-ending emission of greenhouse gases has led to global warming and climate change. Primary sources of greenhouse gas emissions are (https://www.epa.gov/ghgemissions/sources-greenhouse-gas-emissions).

- Transportation.
- Electricity production.
- Industry.
- Commercial and Residential establishments.
- Agriculture Land Use.
- Forestry.

According to EPA (2016), from 1970 to 2011, the burning of fossil fuels and industrial activities contributed about 79% of greenhouse gas emissions. Figure 6.1 shows the total US greenhouse gas emissions by economic sectors in 2019.

The main components in greenhouse gases include carbon dioxide, methane, nitrous oxide, hexafluoroethane, tetrafluoromethane, hydrofluorocarbons, perfluorocarbons, and sulfur hexafluoride. Carbon dioxide is considered as the main contributor to its enormous emission (Meinshausen et al. 2011). Greenhouse gases bring a series of natural calamities which include an increase in the sea level, shrinking glaciers, tropical rain forest decrement, drought, reduction of crop productivity, and ultimately threaten anthropic life. To ease the worldwide greenhouse effect, people are aggressively advocating low carbon lifestyles, circular economy, carbon finance, and other strategies which are sustainable (Zhu et al. 2017).

Carbon dioxide represents 68% of greenhouse gas emission into the atmosphere (Usui and Ikenouchi 1997) and is the main contributor to global warming. The Kyoto Protocol and the Paris Agreement set ambitious goals and responsibilities for participating countries for controlling emissions of greenhouse gases. While these agreements are to limit carbon dioxide emission, there is another aspect in the reduction of carbon dioxide in the atmosphere, i.e., sequestration of carbon dioxide

Fig. 6.1 Total US greenhouse gas emissions by economic sectors in 2019 Total Emissions in 2019 = 6558 *Million Metric Tons of CO₂ equivalent* (https://www.epa.gov/ghgemissions/sources-greenhouse-gas-emissions#colorbox-hidden). Percentages may not add up to 100% due to independent rounding. https://www.epa.gov/ghgemissions/sources-greenhouse-gas-emissions

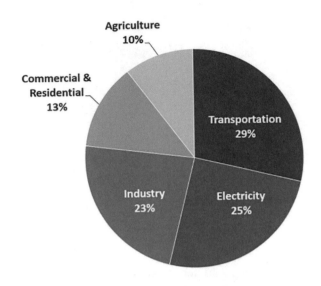

Table 6.1 Benefits of carbon dioxide sequestration

It reduces carbon dioxide concentration in the atmosphere
It reduces or avoids carbon dioxide emission if carbon dioxide is to be captured from large stationary sources
The captured carbon dioxide can be used as a feedstock or substrate for production of chemical and energy products

Based on Zhou et al. (2017), Farrelly et al. (2013); Stepan et al. (2002)

(Giordano et al. 2005; Pires et al. 2012; Kumar et al. 2010; Dah-Wei Tsai et al. 2017). The technical benefits of carbon dioxide sequestration are listed in Table 6.1.

Carbon dioxide sequestration and use also generate new economic and job opportunities in addition to the above technical benefits.

So, it is important to develop a suitable technology for reducing the emissions and accumulation of carbon dioxide.

There are several techniques for carbon dioxide sequestration (White et al. 2003; Hyvonen et al. 2007; Lal 2008; Beedlow et al. 2004; Olajire 2013; Kita and Ohsumi 2004; De Silva et al. 2015; Tang et al. 2014; Salek et al. 2013). These techniques are classified into the physical, chemical, and biological categories (Table 6.2). Each of them has benefits and drawbacks.

Physical storage refers to the processes which directly inject highly concentrated carbon dioxide into the deep ocean, aquifers, or depleted oil/gas wells (Lackner 2003).

Chemical fixation involves carbon dioxide immobilization using adsorption material (such as lithium hydroxide) followed by alkaline-mediated neutralization which leads to the formation of carbonates or bicarbonates. Both have their own benefits and drawbacks (Lackner 2003).

Table 6.2 Carbon dioxide sequestration methods

Method	Mechanisms	Prospects	Limitations
Physical methods			
Membrane separation	Isolation of CO_2 from the main stream by passing mixed gas through a membrane	1. Increased mass transfer	1. Energy inefficient 2. Membrane fouling and blockage 3. High cost
Geologic injection	Injection of CO_2 into geologic reservoirs, depleted oil/gas wells, and coal seams	1. Make use of abandoned space 2. Relatively easy operation 3. Possible recovery of oil/methane	1. Requirement of particular geological and geomorphological environment 2. Gas leakage over time (several thousand years) 3. High cost
Oceanic injection	Injection of CO_2 into deep ocean	1. Large CO_2 holding capacity	1. Gas leakage over time (several hundred years) 2. Threaten the lives of non-swimming marine organisms 3. Requirement of high-cost injection techniques
Adsorption	Using molecular sieves or zeolites	1. Minimal waste generation 2. Flexible to different CO_2 sequestration schemes	1. Energy inefficient 2. Co-adsorption of other components (SOX)
Chemical methods			
Chemical absorption	Neutralization of carbonic acid to form carbonates or bicarbonates	1. Safe and permanent sequestration 2. Rich supply of required base ions (Na+, K+)	1. Large equipment size requirement 2. High energy requirements 3. High cost
Mineral carbonation	Reaction of CO2 with metal oxides to form stable carbonates	1. Abundantly available metal oxides (MgO, CaO) 2. Safe and permanent sequestration 3. Utilization of stable carbonates after sequestration	1. Requirement of large amount of reagent 2. Not cost-effective
Biological methods			
Forestation	Incorporating atmospheric CO_2 into biomass over the lifetime of trees	1. Chemical-free	1. Limited CO_2 sequestration 2. Large land area requirement 3. Potential threat to biological diversity and food supply

(continued)

Table 6.2 (continued)

Method	Mechanisms	Prospects	Limitations
Oceanic fertilization	Triggered growth of photosynthetic organisms by extra iron sources	1. Significant increase in CO_2 sequestration	1. High cost 2. High level of uncertainty 3. Impact on ocean ecosystem (change in plankton structures) 4. possible trigger of methane production
Microalgae-based sequestration	Utilization of CO2 via microalgal photosynthesis	1. High photosynthetic efficiency 2. Efficient in low-concentration CO_2 sequestration 3. Faster sequestration rate than higher plants 4. Do not compete with crops for arable land 5. Co-production of food, feed, fuel, fine chemicals, etc	1. Sensitive to toxic substances in exhaust gases 2. Not very cost-effective for photobioreactors construction and algal biomass foresting

Based on Zhou et al. (2017), White et al. (2003), Hyvonen et al. (2007); Lal (2008), Beedlow et al. (2004), Olajire (2013), Kita and Ohsumi (2004); De Silva et al. (2015), Tang et al. (2014); Salek et al. (2013)

Physical methods like direct carbon dioxide injection are appropriate for large-scale carbon dioxide sequestration. But these methods need certain geological and geomorphological structures, costly separation technologies and equipment for collecting and concentrating carbon dioxide, uncertainties, and danger of long-term leakage, etc. (Lackner 2003).

Chemical neutralization methods are comparatively safer and the long-term carbon dioxide fixation process is not economical, as for neutralization huge amounts of reagents are needed (Lackner 2003). Moreover, both physical and chemical methods are faced with challenges in capturing carbon dioxide from lower concentrations and diffused- or non-point sources (Price and Smith 2008; Nouha et al. 2015).

Microalgal-based approaches can deal with these challenges. Carbon is the major component of microalgal cells. It accounts for about 50% of cell dry weight. There are estimates that 100 tons of microalgal biomass production is equal to about 183 tons of carbon dioxide fixation (Kumar et al. 2010; Sheehan et al. 1998). Microalgae are able to sequester low concentrations of carbon dioxide from air or high-concentration carbon dioxide from stationary sources, for example, coal-burning power plants, and inorganic and organic carbons in wastewater. Moreover, algae can efficiently use nitrogen- and sulfur-containing pollutants, showing the potential of reducing *nitrogen oxides* and sulfur oxides which are potent greenhouse gases (Farrelly et al. 2013; Harun et al. 2010).

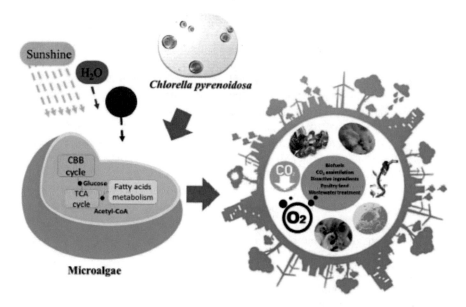

Fig. 6.2 Microalgae for potential biotechnological applications. Reproduced with permission from Zhu et al. (2017)

Terrestrial plants can also sequester vast amounts of carbon dioxide from the atmosphere. But microalgae and cyanobacteria have faster growth rates, and their carbon dioxide fixation efficiency is also between 10 and 50 times higher when compared to terrestrial plants (Costa et al. 2000; Langley et al. 2012). Microalgae appear to be more promising for renewable energy production and contribute not only to developing sustainable energy resources but are able also to protect the environment from air pollution and global warming through sequestration of carbon dioxide (Fig. 6.2) (Zhu et al. 2017).

Microalgae are a class of photoautotropic microorganisms of which there are several groups. Cyanobacteria are another type of photoautotrophs that can be treated as microalgae from an engineering standpoint. However, they are biologically very different from microalgae. Microalgae strains differ widely in a variety of aspects such as growth rate, nutrient needs, or optimal temperature. More importantly, the biomass obtained from different microalgal strains can have very different composition in substances such as protein, pigment, or fatty acid type and amount.

Microalgae have an important ecological role (Khan et al. 2009; Brennan and Owende 2010; Greenwell et al. 2010; Mutanda et al. 2011). Algae are the main producers of oxygen on earth and also serve as food and feed source for people and animals as they belong to the bottom of the food chain (Khan et al. 2009). They can be either autotrophic or heterotrophic, and in some cases even both. Autotrophic microalgae need inorganic compounds, salts, and an appropriate light source for growth, while heterotrophic microalgae use external sources of organic compounds and also nutrients, which are used as an energy source. Microalgae can be classified into two prokaryotic divisions and nine eukaryotic divisions. Microalgae are generally categorized based on their basic cellular structure, pigmentation, and life cycle. As novel genetic and ultra-structural information is continuously emerging, the evolutionary history and taxonomy of microalgae are complex.

Microalgae show promise for the production of value-added products and biofuels. These are rich in minerals, vitamins, oils, and fatty acid methyl esters (Spolaore et al. 2006; Del Campo et al. 2007; Khan et al. 2009; Mutanda et al. 2011). Microalgae can tolerate high concentrations of carbon dioxide, and this inherent ability makes them very beneficial in using carbon dioxide from flue gases of power plants. They grow fast with biomass volumes that double within 24 h.

Farrelly et al. (2013) reported that at a flow rate of 0.3 L/min of air with 4% carbon dioxide concentration, most microalgae are able to achieve a carbon-fixation rate of roughly 14.6 gcm^{-2}/day.

"The biological mitigation of carbon dioxide using microalgae could therefore offer several benefits. No additional carbon dioxide is created, while nutrient utilization is achieved in a continuous fashion leading to the production of biofuels and other secondary metabolites. Therefore, microalgal-mediated carbon dioxide fixation coupled with biofuel production, and wastewater treatment could present a promising alternative to existing carbon dioxide mitigation strategies" (Bhola et al. 2014; Wang et al. 2008; Lam et al. 2012).

"Microalgae usually have a simple life cycle, high growth rate, easy genetic tractability, impressive photosynthetic capability, fertile land independency and high area yield of valuable co-products, these advantages render them attractive feedstock of bioenergy (Malcata 2011; Peng et al. 2016b; Wang et al. 2016). Green algae and cyanobacteria (also called blue-green algae) distribute widely under various ambient, from aquatic to terrestrial some of them even live in extremely arid, thermal, cold or salinity conditions (Kumar et al. 2011). They can convert inorganic carbon (atmospheric carbon dioxide and carbon dioxide in industrial fuel gas) into organic carbon biomass with high chemical energy, such as lipids, carotenoid, acetone and bioethanol as shown in Table 6.3 and Fig. 6.3 (Chow et al. 2015; Jia et al. 2016; Liu et al. 2014; Quintana et al. 2011).

The carbon dioxide sequestration capabilities in photosynthetic and autotrophic microalgae are found 10–50 times more efficient than many higher plants (Cheng et al. 2013). It has been reported that enhancing carbon dioxide concentration would achieve a higher yield of fatty acids in *Chlorella minutissima, Chlamydomonas reinhardtii, Dunaliella tertiolecta, Coccomyxa subellipsoidea C-169*, and *Scenedesmus obliquus* (Farrelly et al. 2014; Morais and Costa 2007; Peng et al. 2016a; Russo et al. 2017; Tang et al. 2011). Understanding the detailed molecular mechanism of carbon dioxide sequestration and assimilation, as well as their regulation network in microalgae might, pave the way to enhance carbon dioxide sequestration and assimilation through the utilization of genetic engineering tools, and finally prompt the productivity of valuable carbon skeleton chemicals, including lipids, glycolate, astaxanthin, bioethanol and so on" (Zhu et al. 2017; Cheng et al. 2017; Chow et al. 2015; Gunther et al. 2012; Nakanishi et al. 2014).

Table 6.3 Representative of important metabolites in green algae and cyanobacteria

Metabolites	Species	Modification designs	Maximum production	References
Glycolate	*C. reinhardtii*	Enhanced the O_2/CO_2-ratio from 30/0.03% to 40/0.024%	13 μmol (mg Chl a^{-1}) h-1	Gunther et al. (2012)
Astaxanthin	*Haematococcus pluvialis*	Mutation of strain by ^{60}Co-y irradiation with 15% CO_2 domestication	1.7-fold higher	Cheng et al. (2017)
Carbohydrate	*Acutodesmus dimorphus*	Addition of 200 mM NaCl for inducement	53.30 ± 2.76%	Chokshi et al. (2017)
Fatty acids	*Chlorella minutissima MCC27*	Co-cultured with *Aspergillus awamori*	3.4–5.2-fold increase	Dash and Banerjee (2017)
1 -butanol	*Synechocystis* sp. PCC6803	Co-overexpressing regulator genes *slr1037* and *5,110,039*	Improved by 133%	Gao et al. (2017)
Bioethanol	*Synechococats elongatus* PCC7942	Co-expressing *icta ecaA* and *acsAB*	7.2 g L^{-1}	Chow et al. (2015)
Intracellular lipids	*Chlamydomonas reinhardtii*	Knockdown *cracsl* and *cracs2*	8.19 mg/109cells	Jia et al. (2016)
Ethylene	*Synechocystis* sp. PCC6803	Introducing codon optimized ethylene-forming enzyme from *Pseudomonas syringae* pv. sesame and used 2-0G as substrate	295 mg L^{-1} day^{-1}	Zhu et al. (2014)
Carotenoids	*Chlorella zofingiensis*	Overexpressing PDS-L516F	Increased 32.1%	Liu et al. (2014)

Reproduced with permission from Zhu et al. (2017)

Sydney (2010) reported that carbon uptake is usually dependent upon the metabolic activity of microalgae. Microalgae exposed to higher levels of carbon dioxide respond better (on a biomass basis), when compared to microalgae exposed to ambient air only. Microalgae can generate about 280 tons of dry biomass per ha per year by using 9% of the freely available solar energy. During this process, about 513 tons of carbon dioxide can be sequestered.

For reducing global warming, microalgae have been studied for many decades as a raw material for renewable energy. These organisms can use concentrated amounts of carbon dioxide, present in power plant flue gases and also from other sources.

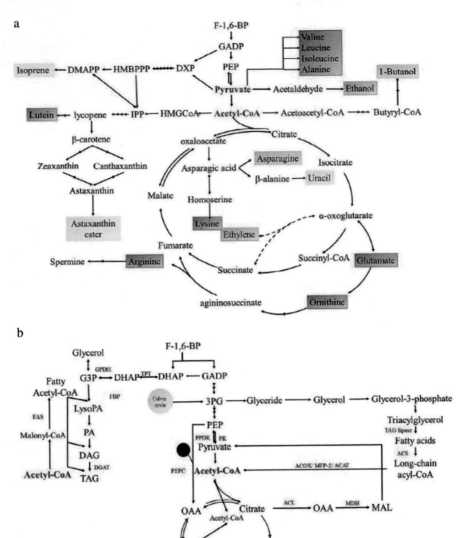

Fig. 6.3 Generation of metabolites through CO_2 sequestration in microalgae. a Metabolic biopro-cess of important metabolites during CO_2 sequestration in microalgae. Various metabolites are presented in different colors. b Fatty acids biosynthesis pathways in microalgae. Reproduced with permission from Zhu et al. (2017)

So, microalgae represent a powerful GHG mitigation strategy. During the middle of the 1970s, the United States Department of Energy started interesting research on microalgal wastewater treatment (Benemann et al. 1977). The recovered microalgal biomass was subjected to anaerobic digestion which produced methane gas. The "Aquatic Species Program" (United States) funded by the office of fuels development started out as a project studying the possibilities of using microalgae for sequestering carbon dioxide emissions from coal power plants (Sheehan et al. 1998). Microalgae that could produce high amounts of oils and also grow under adverse environmental conditions such as extreme temperature, pH, and salinity were screened. This program developed a culture collection system and also supported a pilot project consisting of two raceway ponds in New Mexico. It was suggested that adequate resources would be available in the southwest region of the United States for large-scale production of microalgae capable of capturing several hundred million tons of carbon dioxide annually (Benemann and Oswald 1996). For such a process to be economically feasible, favorable sites which yielded productivities near the theoretical maximum would be needed. During the 1990s in Japan, a major R&D program totaling over $250 million was conducted. The assignment focused on microalgae biofixation of carbon dioxide and also GHG abatement using closed photobioreactors (PBRs). But because of the high costs associated with PBRs, the Japanese initiative did not continue. The use of PBRs was then confined to inoculum production only (Lipinsky 1992; Nakajima and Ueda 2000).

"The United States DOE-NETL promoted microalgae R&D using closed PBRs. Other international participants of R&D pertaining to microalgae carbon dioxide abatement included the following: Arizona Public Services, ENEL Produzione Ricerca, EniTecnologie, ExxonMobil and Rio Tinto. Due to the initiative of these governmental and private industries, the International Network on Microalgae Biofixation of carbon dioxide and GHG Abatement was formed in 2000 in an effort to bring together the limited technical expertise in this field (Pedroni et al. 2001). Past research initiatives suggest that practical carbon dioxide utilization using microalgae still requires innovative scientific and technological breakthroughs to render this a feasible technology. Unless coupled with other technologies or coprocesses, investments into microalgae R&D are unlikely to make a considerable contribution to solving the carbon dioxide problem globally" (Bhola et al. 2014).

The use of microalgae can be classified as a direct carbon dioxide mitigation technology. "Direct strategies usually encompass much higher economic projections, going into billions of dollars, as opposed to indirect approaches. Therefore, for this technology to be a success, future R&D should focus on achieving higher biomass productivities, culture stability over long periods of time, economical harvesting techniques and improved biomass-to-fuels conversion technologies. The economics of microalgal carbon dioxide utilization may be improved by integrating this procedure with other co-processes. Potential co-processes include wastewater treatment, production of useful metabolites, as well as biofuels, animal feed and biofertilizer manufacturing. Wastewater treatment as a co-process has emerged as a viable approach as process requirements and objectives overlap significantly. Municipal wastewater treatment is more favorable when compared to agricultural wastewater treatment as it could yield an animal-feed co-product. This would greatly aid in the economics of the entire process as well as contribute to GHG abatement by not producing additional fossil fuel that is generally required for product formation" (Bhola et al. 2014; Ho et al. 2011).

References

P.A. Beedlow, D.T. Tingey, D.L. Phillips, W.E. Hogsett, D.M. Olszyk, Rising atmospheric CO_2 and carbon sequestration in forests. Front. Ecol. Environ. **2**, 315–322 (2004)

J.R. Benemann, B.L. Koopman, J.C. Weissman, D.M. Eisenberg, W.J. Oswald, Species control in large scale microalgae biomass production. Report to University of California Berkeley SERL 77-5, SAN/740-77/1 (1977)

J.R. Benemann, W.J. Oswald, Systems and economic analysis of microalgae ponds for conversion of CO_2 to biomass. Final report US DOE. http://www.osti.gov/bridge/servlets/purl/493389-FXQ yZ2/webviewable/493389.pdf(1996)

V. Bhola, F. Swalaha, R. Ranjith Kumar, F. Singh, F. Bux, Overview of the potential of microalgae for CO_2 sequestration. Int. J. Environ. Sci. Technol. **11**, 2103–2118 (2014)

L. Brennan, P. Owende, Biofuels from microalgae—A review of technologies for production, processing, and extractions of biofuels and co-products. Renew. Sustain. Energy Rev. **14**, 557–577 (2010)

J. Cheng, Y. Huang, J. Feng, J. Sun, J.H. Zhou, K.F. Cen, Improving CO_2 fixation efficiency by optimizing Chlorella PY-ZU1 culture conditions in sequential bioreactors. Bioresour. Technol. **144**(144C), 321–327 (2013)

J. Cheng, K. Li, Y.X. Zhu, W.J. Yang, J.H. Zhou, K.F. Cen, Transcriptome sequencing and metabolic pathways of astaxanthin accumulated in Haematococcus pluvialis mutant under 15% CO_2. Bioresour. Technol. **228**, 99–105 (2017)

K. Chokshi, I. Pancha, A. Ghosh, S. Mishra, Salinity induced oxidative stress alters the physiological responses and improves the biofuel potential of green microalgae Acutodesmus dimorphus. Bioresour. Technol. **244**, 1376–1383 (2017)

T.J. Chow, H.Y. Su, T.Y. Tsai, H.H. Chou, T.M. Lee, J.S. Chang, Using recombinant cyanobacterium (Synechococcus elongatus) with increased carbohydrate productivity as feedstock for bioethanol production via separate hydrolysis and fermentation process. Bioresour. Technol. **184**, 33–41 (2015)

J.A.V. Costa, G.A. Linde, D.I.P. Atala, Modelling of growth conditions for cyanobacterium Spirulina platensis in microcosms. World J. Microbiol. Biotechnol. **16**, 15–18 (2000)

D. Dah-Wei Tsai, P.H. Chen, R. Ramaraj, The potential of carbon dioxide capture and sequestration with algae. Ecol. Eng. **98**, 17–23 (2017)

A. Dash, R. Banerjee, Enhanced biodiesel production through phyco-myco co-cultivation of Chlorella minutissima and Aspergillus awamori: an integrated approach. Bioresour. Technol. **238**, 502–509 (2017)

G.P.D. De Silva, P.G. Ranjith, M.S.A. Perera, Geochemical aspects of CO_2 sequestration in deep saline aquifers: a review. Fuel **155**, 128–143 (2015)

J.A. Del Campo, M. Garcia-Gonzalez, M.G. Guerrero, Outdoor cultivation of microalgae for carotenoid production: current state and perspectives. Appl. Microbiol. Biotechnol. **74**, 117–1163 (2007)

D.J. Farrelly, C.D. Everard, C.C. Fagan, K.P. McDonnell, Carbon sequestration and the role of biological carbon mitigation: a review. Renew. Sustain. Energy Rev. **21**, 712–727 (2013)

D.J. Farrelly, L. Brennan, C.D. Everard, K.P. McDonnell, Carbon dioxide utilisation of Dunaliella tertiolecta for carbon bio-mitigation in a semicontinuous photobioreactor. Appl. Microbiol. Biotechnol. **98**(7), 3157–3164 (2014)

Y. Gao, T. Sun, L. Wu, L. Chen, W. Zhang, Co-overexpression of response regulator genes slr1037 and sll0039 improves tolerance of Synechocystis sp. PCC 6803 to 1-butanol. Bioresour. Technol. **245**(Pt B):1476–1483 (2017). https://doi.org/10.1016/j.biortech.2017.04.112. Epub 2017 May 4.

M. Giordano, J. Beardall, J.A. Raven, CO_2 concentrating mechanisms in algae: mechanisms, environmental modulation, and evolution. Annu. Rev. Plant Biol. **56**, 99–131 (2005)

H.C. Greenwell, L.M.L. Laurens, R.J. Shields, R.W. Lovitt, K.J. Flynn, Placing microalgae on the biofuels priority list: A review of the technological challenges. J. R. Soc. Interf. **7**, 703–726 (2010)

A. Gunther, T. Jakob, R. Goss, S. Konig, D. Spindler, N. Rabiger, S. John, S. Heithoff, M. Fresewinkel, C. Posten, C. Wilhelm, Methane production from glycolate excreting algae as a new concept in the production of biofuels. Bioresour. Technol. **121**, 454–457 (2012)

R. Harun, M. Singh, G.M. Forde, M.K. Danquah, Bioprocess engineering of microalgae to produce a variety of consumer products. Renew. Sustain. Energy. Rev. **14**, 1037–1047 (2010)

S.H. Ho, C.Y. Chen, D.J. Lee, J.S. Chang, Perspectives on microalgal CO_2-emission mitigation systems—A review. Biotechnol. Adv. **29**, 189–198 (2011)

R. Hyvonen, G.I. Agren, S. Linder, T. Persson, M.F. Cotrufo, A. Ekblad, The likely impact of elevated CO_2, nitrogen deposition, increased temperature and management on carbon sequestration in temperate and boreal forest ecosystems: a literature review. New. Phytol. **2007**(173), 463–480 (2007)

B. Jia, Y. Song, M. Wu, B. Lin, X. Kang, Z. Hu, Y. Huang, Characterization of long-chain acyl-CoA synthetases which stimulate secretion of fatty acids in green algae Chlamydomonas reinhardtii. Biotechnol. Biofuels **9**, 184 (2016)

S.A. Khan, H.M.Z. Rashmi, S. Prasad, U.C. Banerjee, Prospects of biodiesel production from microalgae in India. Renew. Sust. Energy Rev. **13**, 2361–2372 (2009)

J. Kita, T. Ohsumi, Perspectives on biological research for CO_2 ocean sequestration. J. Oceanogr. **60**, 695–703 (2004)

A. Kumar, S. Ergas, X. Yuan, A. Sahu, Q.O. Zhang, J. Dewulf, F.X. Malcata, H. van Langenhove, Enhanced CO_2 fixation and biofuel production via microalgae: recent developments and future directions. Trends Biotechnol. **28**, 371–380 (2010)

K. Kumar, C.N. Dasgupta, B. Nayak, P. Lindblad, D. Das, Development of suitable photobioreactors for CO_2 sequestration addressing global warming using green algae and cyanobacteria. Bioresour. Technol. **102**(8), 4945–4953 (2011)

K.S. Lackner, A guide to CO_2 sequestration. Science **300**, 1677–1678 (2003)

R. Lal, Sequestration of atmospheric CO_2 in global carbon pools. Energy Environ. Sci **2008**(1), 86–100 (2008)

M.K. Lam, K.T. Lee, A.R. Mohamed, Current status and challenges on microalgae-based carbon capture. Int. J. Greenhouse Gas Control **10**, 456–469 (2012)

N.M. Langley, S.T.L. Harrison, R.P. Van Hille, A critical evaluation of CO_2 supplementation to algal systems by direct injection. Biochem. Eng. J. **68**, 70–75 (2012)

E.S. Lipinsky, R&D status of technologies for utilization of carbon dioxide. Energy Convers. Manag. **33**, 505–512 (1992)

J. Liu, Z. Sun, H. Gerken, J.C. Huang, Y. Jiang, F. Chen, Genetic engineering of the green alga Chlorella zofingiensis: a modified norflurazon-resistant phytoene desaturase gene as a dominant selectable marker. Appl. Microbiol. Biotechnol. **98**(11), 5069–5079 (2014)

F.X. Malcata, Microalgae and biofuels: a promising partnership? Trends Biotechnol. **29**(11), 542 (2011)

M. Meinshausen, S.J. Smith, K. Calvin, J.S. Daniel, M.L.T. Kainuma, J.F. Lamarque, K. Matsumoto, S.A. Montzka, S.C.B. Raper, K. Riahi, A. Thomson, G.J.M. Velders, D.P.P. van Vuuren, The RCP greenhouse gas concentrations and their extensions from 1765 to 2300. Clim. Change **109**(1–2), 213–241 (2011)

M.G.D. Morais, J.A.V. Costa, Carbon dioxide fixation by Chlorella kessleri, Cvulgaris, Scenedesmus obliquus and Spirulina sp cultivated in flasks and vertical tubular photobioreactors. Biotechnol. Lett. **29**(9), 1349–1352 (2007)

T. Mutanda, D. Ramesh, S. Karthikeyan, S. Kumari, A. Anandraj, F. Bux, Bioprospecting for hyper-lipid producing microalgal strains for sustainable biofuel production. Bioresour. Technol. **102**, 57–70 (2011)

Y. Nakajima, R. Ueda, The effect of reducing light-harvesting pigment on marine microalgal productivity. J. Appl. Phycol. **12**, 285–290 (2000)

A. Nakanishi, S. Aikawa, S.H. Ho, C.Y. Chen, J.S. Chang, T. Hasunuma, A. Kondo, Development of lipid productivities under different CO_2 conditions of marine microalgae Chlamydomonas sp JSC4. Bioresour. Technol. **152**, 247–252 (2014)

K. Nouha, R.P. John, S. Yan, R. Tyagi, R.Y. Surampalli, T.C. Zhang, Carbon capture and sequestra-tion: biological technologies, in *Carbon Capture and Storage: Physical, Chemical, and Biological Methods* (2015), pp. 65–111

A.A. Olajire, A review of mineral carbonation technology in sequestration of CO_2. J Pet Sci Eng **109**, 364–392 (2013)

P. Pedroni, J. Davison, H. Beckert, P. Bergman, J. Benemann, A proposal to establish an international network on biofixation of CO_2 and greenhouse gas abatement with microalgae. J. Energy Environ. Technol. **1**, 136–215 (2001)

H. Peng, D. Wei, G. Chen, F. Chen, Transcriptome analysis reveals global regulation in response to CO_2 supplementation in oleaginous microalga Coccomyxa subellipsoidea C-169. Biotechnol. Biofuels **9**, 151 (2016a)

H.F. Peng, W. Dong, C. Feng, C. Gu, Regulation of carbon metabolic fluxes in response to CO_2 supplementation in phototrophic Chlorella vulgaris: a cytomic and biochemical study. J. Appl. Phycol. **28**(2), 1–9 (2016b)

J.C.M. Pires, M.C.M. Alvim-Ferraz, F.G. Martins, M. Simoes, Carbon dioxide capture from flue gases using microalgae: engineering aspects and biorefinery concept. Renew. Sustain. Energy Rev. **16**, 3043–3053 (2012)

I. Price, B. Smith, *Carbon Capture and Storage, Meeting the Challenge of Climate Change* (Bluewave Resources LLC of McLean, Virginia, USA, 2008)

N. Quintana, F. Van der Kooy, M.D. Van de Rhee, G.P. Voshol, R. Verpoorte, Renewable energy from Cyanobacteria: energy production optimization by metabolic pathway engineering. Appl. Microbiol. Biotechnol. **91**(3), 471–490 (2011)

D.A. Russo, A.P. Beckerman, J. Pandhal, Competitive growth experiments with a high-lipid Chlamy-domonas reinhardtii mutant strain and its wild-type to predict industrial and ecological risks. AMB Express **7**(1), 10 (2017)

S.S. Salek, R. Kleerebezem, H.M. Jonkers, G.J. Witkamp, M.C.M. van Loosdrecht, Mineral CO_2 sequestration by environmental biotechnological processes. Trends Biotechnol. **31**, 139–146 (2013)

J. Sheehan, T. Dunahay, J. Benemann, P. Roessler, A look back at the U. S. Department of Energy's aquatic species program—Biodiesel from algae. NREL/TP-580-24190 (US Department of Energy's Office of Fuels Development, 1998)

P. Spolaore, C. Joannis-Cassan, E. Duran, A. Isambert, Commercial applications of microalgae. J. Biosci. Bioeng. **101**, 87–96 (2006)

D.J. Stepan, R.E. Shockey, T.A. Moe, R. Dorn, Carbon dioxide sequestering using microalgal systems (University of North Dakota, 2002)

E.B. Sydney, Potential carbon dioxide fixation by industrially important microalgae. Bioresour. Technol **101**, 5892–5896 (2010)

Y. Tang, R.Z. Yang, X.Q. Bian, A review of CO_2 sequestration projects and application in China. Sci World J. Article ID 381854 (2014). https://doi.org/10.1155/2014/381854

H. Tang, M. Chen, M.E. Garcia, N. Abunasser, K.Y. Ng, S.O. Salley, Culture of microalgae Chlorella minutissima for biodiesel feedstock production. Biotechnol. Bioeng. **108**(10), 2280–2287 (2011)

N. Usui, M. Ikenouchi, The biological CO_2 fixation and utilization project by RITE (1) highly-effective photobioreactor system. Energy Convers. Manage. **38**, S487–S492 (1997)

B. Wang, Y.Q. Li, N. Wu, C.Q. Lan, CO_2 bio-mitigation using microalgae. Appl Microbiol. Biotechnol **79**, 707–718 (2008)

X. Wang, W. Liu, C.P. Xin, Y. Zheng, Y.B. Cheng, S. Sun, R. Li, X.G. Zhu, S.Y. Dai, P.M. Rentzepis, J.S. Yuan, Enhanced limonene production in cyanobacteria reveals photosynthesis limitations. Proc. Natl. Acad. Sci. u.s.a. **113**(50), 14225–14230 (2016)

C.M. White, B.R. Strazisar, E.J. Granite, J.S. Hoffman, H.W. Pennline, Separation and capture of CO_2 from large stationary sources and sequestration in geological formations - coalbeds and deep saline aquifers. J. Air. Waste Manage. Assoc. **53**, 645–715 (2003)

W. Zhou, J. Wang, P. Chen, C. Jia, Q. Kanga, B. Lua, K. Lia, J. Liud, R. Ruan, Bio-mitigation of carbon dioxide using microalgal systems: Advances and perspectives. Renew. Sustain. Energy Rev. **76**, 1163–1175 (2017)

B. Zhu, G. Chen, X. Cao, D. Wei, Molecular characterization of CO_2 sequestration and assimilation in microalgae and its biotechnological applications. Biores. Technol. **244**, 1207–1215 (2017)

T. Zhu, X. Xie, Z. Li, X. Tan, X. Lu, Enhancing photosynthetic production of ethylene in genetically engineered *Synechocystis* sp. PCC 6803. Green Chem. **17**(1), 421–434 (2014)

Chapter 7
Water Footprint

"The water footprint is a measure of humanity's appropriation of fresh water in volumes of water consumed and/or polluted". It can be measured for a single process, or for a product.

"ISO 14046:2014 specifies principles, requirements, and guidelines related to water footprint assessment of products, processes, and organizations based on a life cycle assessment.

A water footprint assessment can assist in.

- Evaluating the extent of potential environmental impacts related to water.
- Recognizing opportunities for reducing water related potential environmental impacts associated with products at different stages in their life cycle and also processes and organizations.
- Strategic risk management related to water.
- Facilitating water efficiency and optimization of water management at process, product, and organizational levels.
- Providing information to decision-makers in industry, government or nongovernmental organizations of their potential environmental impacts related to water (example for the purpose of strategic planning, priority setting, product or process design or redesign and decisions about investment of resources).
- Providing consistent and reliable information, based on scientific evidence for reporting water footprint results.

A water footprint assessment alone is not sufficient to be used for describing the overall potential environmental impacts of products, processes, or organizations. The water footprint assessment according to this International Standard can be conducted and reported as a standalone assessment where only impacts related to water are assessed, or as part of a life cycle assessment where consideration is given to a comprehensive set of environmental impacts and not only impacts related to water. In this International Standard, the term *water footprint* is only used when it is the

P. Bajpai, *Fourth Generation Biofuels*,
SpringerBriefs in Applied Sciences and Technology,
https://doi.org/10.1007/978-981-19-2001-1_7

result of an impact assessment. The specific scope of the water footprint assessment is defined by the users of this International Standard in accordance with its requirements. Polluted water can be expressed in m^3 per product unit" (Biron 2020).

Hoekstra (2003) first introduced the concept of water footprint. This led to the idea of sustainable water consumption over the full supply chain (Patzelt et al. 2015). This concept is helpful in addressing these issues and relating the production of terrestrial and aquaculture crops to water pollution and scarcity. Water footprint evaluates the water consumed by its source, the amount of polluted water, and the type of pollution (Aldaya et al. 2012).

Three independent components of water footprints (green, blue, and gray) determine the water consumption involved in the cultivation of algae (Gerbens-Leenes 2018). The consumed volumes of rainwater and groundwater in the production process are referred to as green and blue water footprints, respectively. **Gray water footprint** is the "amount of freshwater needed to assimilate pollutants for meeting specific water quality standards. The grey water footprint considers point-source pollution discharged to a freshwater resource directly through a pipe or indirectly through runoff or leaching from the soil, impervious surfaces, or other diffuse sources" (Gerbens-Leenes et al. 2013).

Zhang et al. (2014a) presented life cycle water footprints of biofuels from biomass in China based on the resource distribution, climatic conditions, soil conditions, and crop growing characteristics. Life cycle water footprints including blue, green, and gray water were examined for the selected fuel pathways. Geographical differences in water requirements were shown to be different by location. Water irrigation requirements were considerably different from crop to crop, ranging 2–293, 78–137, and 17–621 m^3/ha, for sweet sorghum, cassava, and Jatropha curcas L., respectively. Four biofuel pathways were selected on this basis to analyze the life cycle water footprint:sweet sorghum-based bioethanol in Northeast China, cassava-based bioethanol in Guangxi, Jatropha curcas L.-based biodiesel in Yunnan, and microalgal-based biodiesel in Hainan. The life cycle water footprints of bioethanol from cassava and sweet sorghum were 3708 and 17,156 m^3 per ton of bioethanol, respectively, whereas those for biodiesel produced from *Jatropha curcas* L. and microalgae were 5787 and 31,361 m^3 per ton of biodiesel, respectively. The crop-growing stage was the major contributor to the whole life cycle of each pathway. In comparison to blue and green water, gray water was noteworthy because of the use of fertilizer during the growing of biomass. From the standpoint of the water footprint, cassava-based bioethanol in Guangxi and Jatropha-based biodiesel in Yunnan were appropriate for promotion, whereas the promotion for microalgae-based biodiesel in Hainan needed improvement on technology. Direct green water footprint of the microalgae was almost a quarter of the average direct green water footprint of the three terrestrial plants.

For algae production, freshwater, seawater, and wastewater can be used (Salama et al. 2017). The use of undiluted wastewater is a good substrate for growing microalgae because it contains a high level of nutrients and also it saves freshwater resources.

De Francisci et al. (2018) studied the production of microalgae coupled with wastewater treatment. It was found that the use of wastewater for growing algae

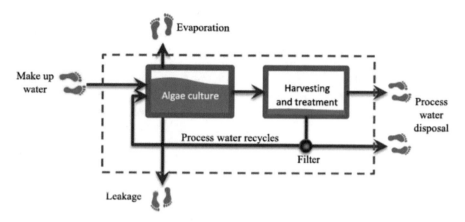

Fig. 7.1 Schematic representation of the Water Footprint and Water demand in genetically modified algae biomass utilization. Abdullah et al. (2019). Reproduced with permission

can make the production of biofuel economically attractive and environmentally sustainable.

Open-pond cultivation has a higher footprint than contained systems. Subhadra and Edwards (2011) reported that the water footprint for microalgae production in an open pond can reach 5.5 ha-ft/acre/year.

Yang et al. (2011) quantified the water footprint of the production of biodiesel from microalgae. For producing 1 kg biodiesel, 3726 kg water, 0.71 kg phosphate, and 0.33 kg nitrogen are needed if freshwater is used without recycling. Recycling the harvested water reduces water and the use of nutrients by 84 and 55%. The use of seawater/wastewater reduces the requirement for water by 90%. The use of nutrients except phosphate is also eliminated.

Schematic water footprint and water demand associated with the production of algal biofuel is shown in Fig. 7.1.

In the biofuel production process, harvesting is the most important phase. There are two major potential choices regarding the water footprint of algal biofuel: Culture with recycling and Culture with the disposal of the harvested water.

Recycling the discharged water from the harvesting phase can reduce the amount of water consumed during the production of biofuel by up to 90.2% (Feng et al. 2016).

The water footprint of biofuels from microalgal biomass has two components of gray and blue waters. When all the water discharged from the algae culture medium is recycled using the best practices, it is presumed that there would be no more water pollution, and the amount of gray water footprint would be zero.

Recycling wastewater from the harvesting stage of genetically modified algae reduces the requirement for fresh media (Gonzalez-Gonzalez et al. 2018). Using a nutrient-recycling system in large-scale microalgae production could improve the efficiency of algal biofuel production in terms of saving blue water, increasing economic opportunities, and reducing material input (Lowrey et al. 2016).

Anaerobic digestion technology has been widely studied for nutrient recycling in algae exploitation. This method can be used for large-scale recycling of different types of organic waste (Sialve et al. 2009).

Crofcheck and Crocker (2016) examined the effect of recycling media and the use of mineralized nutrients during cultivation of *Scenedesmus*. The recycled media was found to support cell growth with nutrient replacement, and could be recycled for cultivation up to four times. Algal biomass was subjected to anaerobic digestion, and the liquid digestate and the total digestate were examined as sources of nutrients. The digestate was rich in ammonium ions and was found to be a satisfactory replacement for urea. When both ammonium and urea ions were present in the media, the assimilation of urea by algal cells was found to be reduced as compared to the case where urea was used as the nitrogen source.

Barbera et al. (2016) used the aqueous phase obtained from flash hydrolysis of *Scenedesmus* sp. as a growth medium for a microalga of the same genus, for assessing the feasibility of this method for nutrient recycling. Batch and continuous cultivations were conducted for determining the growth performances in this substrate in comparison to the standard media, and to confirm if a stable biomass production could be achieved. In continuous culture, the effect of hydrolysate inlet concentration and residence time were studied for optimizing nutrient supply in relation to productivity. Nutrient recycling was found to be feasible by treating biomass with flash hydrolysis. *Scenedesmus* is capable of recycling large amounts of recovered nutrients.

Nutrient recycling from algal biomass residue appears to be important for reducing the environmental impacts and the cost associated with fuels obtained from algae. Zhang et al. (2014b) selected anaerobic digestion and hydrothermal liquefaction and made a comparison of their nutrient recycling and energy recovery potential for lipid-extracted algal biomass using *Scenedesmus dimorphus*. For 1 kg (dry weight) of algae grown in an open raceway pond, 3.8 g phosphorous and 40.7 g nitrogen can be recycled through anaerobic digestion, whereas 6.8 g phosphorous and 26.0 g nitrogen can be recycled through hydrothermal liquefaction. In terms of energy production 2.49 MJ heat and 2.61 MJ electricity are generated from anaerobic digestion biogas combustion to meet production system demands, whereas 0.95 MJ electricity and 3.30 MJ heat from hydrothermal liquefaction products are produced and used within the production system. Assuming recycled nutrient products from anaerobic digestion or hydrothermal liquefaction techniques displace demand for synthetic fertilizers, and energy products displace natural gas and electricity, the life cycle greenhouse gas reduction obtained by adding anaerobic digestion to the simulated algal oil production system is between 622 and 808 g carbon dioxide equivalent/kg biomass depending on substitution assumptions, while the life cycle greenhouse gas reduction achieved by hydrothermal liquefaction is between 513 and 535 g carbon dioxide equivalent/kg biomass depending on substitution assumptions. Based on the efficiency of nutrient recycling and energy recovery, and also technology maturity, anaerobic digestion performs better than hydrothermal liquefaction as a nutrient and energy recycling technology in the production of algal oil.

Talbot (2015) studied the use of hydrolysate obtained after Flash hydrolysis of *Scenedesmus* at 280°C as a nutrient source for growing microalgae. Comparison of flash hydrolysis hydrolysate nutrient recycling was done with low-temperature batch hydrothermal liquefaction nutrient recycling. *Scenedesmus* and *Oocystis* were grown using hydrolysate as a partial nitrogen and phosphorous source. It was found that 50% of the phosphorus needed in the culture media could be replaced with hydrolysate from flash hydrolysis; also, 50% nitrogen was provided from the same source. Ammonia toxicity was the one limitation for the higher percentages of nitrogen replacement in this study. The hydrothermal liquefaction aqueous phase had a near-continuous high soluble ammonia concentration in the media. The effect was important for the 50% phosphorus hydrothermal liquefaction replacement treatment where almost no algal growth was seen during the first 11 days of the experiment.

The disposal of genetically modified algae is a matter of concern in the production of fourth-generation biofuel. The residual water from growing algae cannot be disposed of in the environment without further treatment. Wastewater generated from the harvesting and dewatering steps is channeled through a drainage system into the treatment sump which is disposed of following treatment. Diverse methods are used to remove the inhibitory and remediation of aquaculture in genetically modified microalgal growth. These methods include ultraviolet treatment, chemical deactivation, heat deactivation, pasteurization, dilution, and filtration (Patzelt et al. 2015; Beacham et al. 2017).

Patzelt et al. (2015) used ultraviolet, carbon filtration, and dilution methods to remediate the residual medium of algal growth. Hydrothermal gasification is a process which uses any biomass or carbon-containing source as substrate for producing biogas. Patzelt et al. (2015) studied the potential of using the residual water of the conversion process as a recycled nutrient source for growing microalgae. "Nutrient recycling was studied by checking the growth of *Acutodesmus obliquus* and *Chlorella vulgaris* on residual water from hydrothermal gasification of *A. obliquus*. Four different gasification setups were studied. After the procedure, all obtained liquid nutrient phases contained, besides nutrients, growth-inhibiting substances affecting the photosynthetic activity and biomass yield of the two algal species. At least 28 potential toxic substances were found within one of the batches. Phytotoxicity on the cellular structure was confirmed by electron microscopy. The cell form remained intact but cell compartments disappeared. *C. vulgaris* was not able to recover to a vital growing organism during growth, whereas *A. obliquus* was able to restore cell compartments, photosynthetic activity, and growth after 3 days of growth. A 355-fold dilution, ultraviolet treatment for 4 h and activated carbon filtration of the residual water from gasification finally enabled the discharge to support microalgal growth. Ultraviolet treatment removed 23 substances but produced 4 new substances which were not noticed before treatment. Activated carbon filtration eliminated 26 substances. Growth of microalgae obtained in the treated residual water was comparable with that in the control medium. This study showed the possibility to recover nutrients after the hydrothermal gasification process when the discharge got remediated to restart the value-adding chain of microalgae and reduce additional nutrient supply for growing microalgae".

The load of detrimental substances in an aqueous phase was reduced by using photodegradation of ultraviolet radiation (Hanelt et al. 2006). Of 28 potential toxic compounds, 23 were removed via the irradiation of ultraviolet light. Furthermore, four new compounds were produced during the treatment process. Twenty-six out of 28 toxic substances were removed during carbon filtration (Macova et al. 2010).

The use of algal biofuel has a lower water footprint when compared with other biofuel substrates. The reuse of the discharged water from the harvesting process offers an additional advantage in terms of water footprint. But the culture medium can be recycled a limited number of times and must be disposed of after reaching the limit. Discharging the culture medium from growing genetically modified algae can pose environmental and health risks. Specific water treatment methods must be devised before wastewater can be discharged safely.

References

B. Abdullah, S.A.F. Syed Muhammad, Z. Shokravi, S. Ismail, K.A. Kassim, A.N. Mahmood, M.M.A. Aziz, Fourth generation biofuel: A review on risks and mitigation strategies. Renew. Sustain. Energy Rev. **107**, 37–50 (2019)

M.M. Aldaya, A.K. Chapagain, A.Y. Hoekstra, M.M. Mekonnen, *The Water Footprint Assessment Manual: Setting the Global Standard* (Routledge, 2012)

E. Barbera, E. Sforza, S. Kumar, T. Morosinotto, A. Bertucco, Cultivation of *Scenedesmus obliquus* in liquid hydrolysate from flash hydrolysis for nutrient recycling. Bioresour Technol **207**, 59–66 (2016)

M. Biron, Metrics of Sustainability in plastics: indicators, standards, software, in *A Practical Guide to Plastics Sustainability* (Elsevier, 2020)

T.A. Beacham, J.B. Sweet, M.J. Allen, Large scale cultivation of genetically modified microalgae: a new era for environmental risk assessment. Algal. Res. **25**, 90–100 (2017)

C. Crofcheck, M. Crocker, Application of recycled media and algae-based anaerobic digestate in Scenedesmus cultivation. J. Renew. Sustain. Energy **8**, 1 (2016)

D. De Francisci, Y. Su, A. Iital, I. Angelidaki, Evaluation of microalgae production coupled with wastewater treatment. Environ. Technol. **39**, 581–592 (2018)

P.-Z. Feng, L.-D. Zhu, X.-X. Qin, Z.-H. Li, Water footprint of biodiesel production from microalgae cultivated in photobioreactors. J. Environ. Eng. **142**, 4016067 (2016)

P.W. Gerbens-Leenes, Green, blue and grey bioenergy water footprints, a comparison of feedstocks for bioenergy supply in 2040. Environ. Process 1–14 (2018)

P.W. Gerbens-Leenes, L. de Vries, L. Xu, The water footprint of biofuels from microalgae. Bioenergy Water 191 (2013)

L.M. Gonzalez-Gonzalez, L. Zhou, S. Astals, S.R. Thomas-Hall, E. Eltanahy, S. Pratt, Biogas production coupled to repeat microalgae cultivation using a closed nutrient loop. Bioresour. Technol. **2018**(263), 625–630 (2018)

D. Hanelt, I. Hawes, R. Rae, Reduction of UV-B radiation causes an enhancement of photoinhibition in high light stressed aquatic plants from New Zealand lakes. J. Photochem. Photobiol. B Biol. **84**, 89–102 (2006)

A.Y. Hoekstra, Virtual water trade, in *Proceedings of the International Expert Meeting on Virtual Water Trade* (IHE Delft, The Netherlands, 2003)

J. Lowrey, M.S. Brooks, R.E. Armenta, Nutrient recycling of lipid-extracted waste in the production of an oleaginous thraustochytrid. Appl. Microbiol. Biotechnol. **100**, 4711–4721 (2016)

M. Macova, B.I. Escher, J. Reungoat, S. Carswell, K.L. Chue, J. Keller, Monitoring the biological activity of micropollutants during advanced wastewater treatment with ozonation and activated carbon filtration. Water Res. **2010**(44), 477–492 (2010)

D.J. Patzelt, S. Hindersin, S. Elsayed, N. Boukis, M. Kerner, D. Hanelt, Hydrothermal gasification of Acutodesmus obliquus for renewable energy production and nutrient recycling of microalgal mass cultures. J. Appl. Phycol. **27**, 2239–2250 (2015)

E.S. Salama, M.B. Kurade, R.A.I.. Abou-Shanab, M.M. El-Dalatony, I.S. Yang, B. Min, Recent progress in microalgal biomass production coupled with wastewater treatment for biofuel generation. Renew. Sustain. Energy Rev. **79**, 1189–211

B. Sialve, N. Bernet, O. Bernard, Anaerobic digestion of microalgae as a necessary step to make microalgal biodiesel sustainable. Biotechnol. Adv. **27**, 409–416 (2009)

B.G. Subhadra, M. Edwards, Coproduct market analysis and water footprint of simulated commercial algal biorefineries. Appl. Energy **88**, 3515–3523 (2011)

C.R. Talbot, Comparing nutrient recovery via rapid (flash hydrolysis) and conventional hydrothermal liquefaction processes for microalgae cultivation (2015)

J. Yang, M. Xu, X. Zhang, Q. Hu, M. Sommerfeld, Y. Chen, Life-cycle analysis on biodiesel production from microalgae: water footprint and nutrients balance. Bioresour. Technol. **102**, 159–165 (2011)

T. Zhang, X. Xie, Z. Huang, Life cycle water footprints of nonfood biomass fuels in China. Environ. Sci. Technol. **48**, 4137–4144 (2014a)

Y. Zhang, A. Kendall, J. Yuan, A comparison of on-site nutrient and energy recycling technologies in algal oil production. Resour. Conserv. Recycl. **88**, 13–20 (2014b)

Chapter 8
Future Prospects and Key Challenges

"With the advantages of large-volume biomass and non-competitive with crops, microalgae bring hopes for resolving the problem of bioenergy supply" (Su et al. 2017). Algae appear to be an attractive source for biofuel production. Bioethanol, biodiesel, biogas, nutraceuticals, pharmaceuticals, and several other important products can be obtained from algae. Biofuels are sustainable, biodegradable, and environmentally friendly and remain the most practical solution to the global fuel crisis. Algae show several desirable characteristics such as faster growth and higher lipid content.

The last decade has seen huge investment in liquid biofuel alternatives to petroleum and diesel, with the United States, Brazil, and the European Union leading the way. Many demonstration projects have been launched worldwide for assessing the technologies developed in the laboratory with the support of a financial group and government investment. While the algal biofuels research and development has achieved technological advancement, the commercialization of algal biofuels still cannot be cost-competitive with petroleum. Governments of several countries have made huge investments and scheduled projects with the aim of promoting technology development and reducing the production cost. "With the multi-supports of investment from fiscal departments, large energy companies and venture investment companies, many algae bioenergy companies come into the stage of pivot-scale and industrial demonstration. However, there is still much work to do to achieve commercialization, indeed, some companies shifted to other fields because they cannot realize commercialization. Bio-refinement is the major mode of production in current algae biofuel enterprises. Algae fuel cost can be reduced by producing high value-added products such as nourishment and cosmetics. The shift of the algae company can help them to survive first. As the progress of technology, they could come back to biofuel pilot test and industrialization" (Su et al. 2017).

In general, biofuels have no or very little negative impact on the environment (Adeniyi et al. 2018).

P. Bajpai, *Fourth Generation Biofuels*,
SpringerBriefs in Applied Sciences and Technology,
https://doi.org/10.1007/978-981-19-2001-1_8

Biofuel production process could be associated with other environmental applications including bio-remediation, biofixation, biofertilizer, food, animal feed, healthcare products, and electricity or heat production. Algae are the most sustainable fuel feedstock which may help in reducing emissions of greenhouse gasses (Suganya et al. 2016). As emission of carbon dioxide from liquid fuels was 36% in 2012 (Slade and Bauen 2013; Saladini et al. 2016) which may increase up to 45,000 mega tons by 2040 (Bhore 2014), the European Union Renewable Energy Directive (RED) recommended producing up to 15% of energy from sustainable sources (Saladini et al. 2016) with an objective to significantly reduce emissions of greenhouse gas to 20% by 2050 (Tomei and Helliwell 2016). The United States proposed replacing 20% of its road transport fuel with biofuel by 2022 and a Renewable Transport Fuel Certificate (RTFC) would be granted to fulfill this requirement (Wise and Cole 2015). It is expected that by the year 2070, renewable energy will dominate (Adeniyi et al. 2018).

"The future of algal biofuels depend on the establishment of cost-effective technologies for commercialization. However, the most likable solution is to use genetically-engineered algae with precursor overproduction rates and faster growth rates. These species can be grown in an open pond which is closer to a polluted area. These open ponds could be the first step of water purification in a wastewater plant. The biomass slurry from this pond can be used to produce biofuels and the spent biomass can be used as animal feed or fertilizers. Algal biofuels might seem unsustainable and too expensive, and their production needs a lot of water, nitrogen, phosphorous, and CO_2 but they are environment friendly with no competition on land or water resources" (Saad et al. 2019).

"Large-scale cultures for biofuel production need a lot of tools, equipment, energy, water and nutrients for each step separately. Large-scale algal cultures and large-scale agriculture could use the same amount of nutrients. Generally, algal biofuels possess no or very little negative impact on the environment" (Saad et al. 2019; Adeniyi et al. 2018; Office of Energy Efficiency and Renewable Energy 2010).

Replacement of 5% of gasoline/diesel with algal biofuels might need about 123 billion to 143 trillion liters of water. The use of wastewater avoids the competition with food and feed crops, but also provides some nitrogen and phosphorous. But the main disadvantage is that wastewater may contain some algal pathogens and predators, heavy metals, and other contaminants. Furthermore, most of the algae production sites are far away from wastewater treatment plants rendering them not feasible, expensive, and energy-intensive for carrying low-quality water for long distances. Besides, water will not be completely recycled (Office of Energy Efficiency and Renewable Energy 2010). Coupling of wastewater treatment with biofuel production appears to be the most promising scenario (Cabanelas et al. 2013). Algae require about 14–21 kg of carbon dioxide (Office of Energy Efficiency and Renewable Energy2010, $6–15 \times 10^9$ kg of nitrogen, and $1–2 \times 10^9$ kg of phosphorus per gallon of biodiesel (Pimentel and Patzek 2015). Many industrial sources of carbon dioxide contain heavy metals such as lead, mercury, cadmium, and arsenic. Algae are exceptional at absorbing these toxic elements, but the residue cannot be used as animal feed. Interestingly, the land needed for culturing sufficient algal biomass to replace around 5% of petroleum (which is equivalent to 10 billion gallons) in the United Stated is about 1.5 million acres in the United States Southwest, 2.2–2.4 million acres in Georgia/Florida and United States Midwest, and 4.8 million acres in Texas (Office of Energy Efficiency and Renewable Energy 2010).

There are several economic and technical challenges associated with the use of microalgae in the biofuels industry. Harvesting of microalgae appears to be the main problem. The unicellular algae which stores lipids have lower densities and are found in suspensions making separations quite difficult. "The extraction processes used for large-scale production are particularly complex and are still in the early development stages. Microalgae grown in open pond systems are susceptible to contamination. Bacterial contamination aggressively competes for nutrients and oxidises the organic matter, which can lead to culture putrefaction. They are also susceptible to protozoa and zooplankton grazers that consume microalgae and may destroy the concentrations of algae in a short time. In open pond systems there is also loss of water through evaporation and in order to maintain a fixed volume and salinity in the culture it is important to add large quantities of freshwater. Other challenges that inhibit the commercialization of algal based biofuel production include; difficulties in finding rapid growing algae strains with high oil content, photosynthetic efficiency, simple algae culture harvesting systems, infrastructure, operation and maintenance costs and the ability to develop economical photo-bioreactor designs" (Bajpai 2019; Rawat et al. 2013; Das et al. 2015; Adenle et al. 2013).

According to Saad et al. (2019), "the major challenges appear to be the high infrastructure, operation, and maintenance costs, selection of high lipid containing algal strains, harvesting on a commercial scale and water evaporation issues. Innovative and efficient techniques are necessary to make algal biofuel production preferable. Enhanced biofuel production will help in natural resources conservation and in turn saving the environment".

References

O.M. Adeniyi, U. Azimov, A. Burluka, Algae biofuel: Current status and future applications. Renew. Sustain. Energy Rev. **90**, 316–335 (2018)

A.A. Adenle, G.E. Haslam, L. Lee, Global assessment of research and development for algae biofuel production and its potential role for sustainable development in developing countries. Energy Policy **61**, 182–195 (2013)

P. Bajpai, *Third Generation Biofuels* SpringerBriefs in Energy, Springer Singapore (2019). https://doi.org/10.1007/978-981-13-2378-2_8

N. Bhore, *Energy Outlook: A View to 2040. Detroit Automotive Petroleum Forum* (Detroit, MI, USA, 2014). Available online https://www.api.org/~/media/files/certification/engine-oil-diesel/forms/whats-new/6-energy-outlook-view%20to%202040-nbhore-exxonmobil.pdf

I.T.D. Cabanelas, J. Ruiz, Z. Arbib, F.A. Chinalia, C. Garrido-Pérez, F. Rogalla, I.A. Nascimento, J.A. Perales, Comparing the use of different domestic wastewaters for coupling microalgal production and nutrient removal. Bioresour. Technol. **131**, 429–436 (2013)

P. Das, M.I. Thaher, M.A.Q.M.A. Hakim, H.M.S.J. Al-Jabri, Sustainable production of toxin free marine microalgae biomass as fish feed in large scale open system in the Qatari desert. Bioresour. Technol. **192**, 97–104 (2015)

Office of Energy Efficiency and Renewable Energy, National Algal Biofuels Technology Roadmap; U.S. Department of Energy: Washington, DC, USA (2010)

D. Pimentel, T.W. Patzek, Ethanol production using corn, switchgrass, and wood; Biodiesel production using soybean and sunflower. Nat. Resour. Res. **14**, 65–76 (2015)

I. Rawat, R.R. Kumar, T. Mutanda, F. Bux, Biodiesel from microalgae: a critical evaluation from laboratory to large scale production. Appl Energy **103**, 444–467 (2013)

F. Saladini, N. Patrizi, F.M. Pulselli, N. Marchettini, S. Bastianoni, Guidelines for emergy evaluation of first, second and third generation biofuels. Renew. Sustain. Energy Rev. **66**, 221–227 (2016)

M.G. Saad, N.S. Dosoky, M.S. Zoromba, H.M. Shafik, Algal biofuels: current status and key challenges. Energies **12**, 1920 (2019). https://doi.org/10.3390/en1210192

R. Slade, A. Bauen, Micro-algae cultivation for biofuels: cost, energy balance, environmental impacts and future prospects. Biomass Bioenergy **2013**(53), 29–38 (2013)

Y. Su, K Song, P. Zhang, Y. Su, J. Cheng, X. Chen, Progress of microalgae biofuel's commercialization. Renew. Sustain. Energy Rev. **74**(C), 402–411 (2017)

T. Suganya, M. Varman, H.H. Masjuki, S. Renganathan, Macroalgae and microalgae as a potential source for commercial applications along with biofuels production: a biorefinery approach. Renew. Sustain. Energy Rev. **55**, 909–941 (2016)

J. Tomei, R. Helliwell, Food versus fuel? Going beyond Biofuels. Land Use Policy **2016**(56), 320–326 (2016)

T.A. Wise, E. Cole, *Mandating Food Insecurity: The Global Impacts of Rising Biofuel Mandates and Targets; Global Development and Environment Institute: Medford* (MA, USA, 2015), p. 2015

Index

© The Author(s), under exclusive license to Springer Nature Singapore Pte Ltd. 2022
P. Bajpai, *Fourth Generation Biofuels*,
SpringerBriefs in Applied Sciences and Technology,
https://doi.org/10.1007/978-981-19-2001-1

Printed in the United States
by Baker & Taylor Publisher Services